BRIAN PARKINSON

Tim Lovegrove
TECHNICAL EDITOR

Ki enei kirehe ataahua e whakamiharo ana i tenei tiritiri o te moana, ki nga manu a Tangaroa, me nga tangata kei te mahi, kia rokiroki ai aua manu.

Dedicated to the seabirds, those wonderful creatures that add so much to these southern climes, and to the people who work to preserve them.

First published in 2000 by
New Holland Publishers (NZ) Ltd
Auckland • Sydney • London • Cape Town

218 Lake Road, Northcote, Auckland, New Zealand
14 Aquatic Drive, Frenchs Forest, NSW 2086, Australia
24 Nutford Place, London W1H 6DQ, United Kingdom
80 McKenzie Street, Cape Town 8001, South Africa

ISBN: 1 877246 32 8

Managing editor: Renée Lang
Cover design: Gerrard Malcolm
Design: Gerrard Malcolm
Project editor: Brian O'Flaherty
Technical editor: Tim Lovegrove
Bird topography: Richard Allen
Colour reproduction by Colour Symphony Pte Ltd
Printed by Kyodo Printing Co (Singapore) Pte Ltd
Front cover and introduction photographs: © Dennis Buurman
Back cover and contents page photographs: (top to bottom) © Tim Lovegrove, Brian Parkinson, Dennis Buurman, DOC/M. Williams

Contents

Introduction

© Dennis Buurman

New Zealand: a centre of seabird activity

New Zealand's well-developed tourism infrastructure makes it easy for the seabird enthusiast to travel and explore.

New Zealand is one of the major seabird centres of the world. Because of the country's extensive coastline and many inshore and offshore islands, it is possible for the keen observer to see more seabirds here than in most other countries. Three-quarters of the world's albatrosses, penguins and petrels and half the shearwaters and shags are to be found here, along with numerous representatives of a number of other groups. Moreover, a good proportion of these can be viewed without going far offshore.

New Zealand's well-developed transport and tourism infrastructure means it is reasonably easy for the seabird enthusiast to travel and explore. Also, there is a growing number of ecotourism operators, including several that cater specifically for the seabird enthusiast, operating out of North, South and Stewart Island ports.

Field Guide to New Zealand Seabirds covers the species that are most likely to be encountered in New Zealand waters. Each species is illustrated with a photograph, and is described in terms of key identification and behavioural characteristics, similar species, distribution and breeding areas, and population status. Advice on the best places to see the species is also given.

Recent taxonomic revisions have resulted in numerous name changes, making some previously well-known common names invalid. Readers should refer to the scientific names to confirm which species is being described.

Some species, particularly the albatrosses and gulls, can vary considerably according to age. For space considerations, it is not practical to show all of these plumages. Instead a typical adult, and in some cases a juvenile, are featured.

Seabirds and people

Before the arrival of humans, mainland New Zealand was a dramatically different place. It was a land dominated by large populations of birds, living in an environment free of any mammalian predators. In the mountains, forests, wetlands, open country and coasts, the air would have been filled with their calls. Many early settlers, both Maori and European, remarked on the sheer volume and beauty of the 'Mara o Tane' or 'dawn chorus'.

And if the daytime was noisy, the nights would also have been filled with sound. Although there were nocturnal landbird species such as Kiwi, Morepork, Laughing Owl and Owlet-nightjars, the night would have been largely given over to a cacophony of seabird calls. Countless millions once nested inland, often high in mountain ranges great distances from the sea.

The arrival of humans quickly altered this scene. Human impacts on New Zealand seabird populations have been devastating for the majority of species. Birds were a major food source for the new inhabitants. Conspicuous and vulnerable seabirds like the native Pelican and the Merganser, both now extinct, were probably immediately targeted by Maori settlers. Others, like shearwaters, petrels, shags and gulls, were often harvested annually. Shag chicks were taken from nesting sites on pari kawau or 'shag cliffs'. Shearwaters and petrels on their flight inland to their nesting burrows were attracted by fires lit in front of large nets in mountain passes; these sites were known as ahi-titi or 'shearwater fires'.

By the use of rahui or reservations, some equilibrium may have been established. However, it was the damage inflicted by animals introduced by humans that was to be the most destructive. When Maori arrived in New Zealand they brought with them the Kiore or Pacific Rat (*Rattus exulans*) and the now extinct Kuri (dog); the Kiore in particular was to have a major impact on the biota. It was once thought that the Kiore's effect on the fauna was relatively minor. However, recently it has been determined that many species of invertebrates, along with various species of snails, lizards and frogs, were exterminated by Kiore soon after this rat arrived in New Zealand. The smaller landbirds, and seabirds such as the smaller petrels, the storm petrels and prions were also affected.

© Brian Parkinson

Weka introduced to various islands have killed large numbers of seabirds.

The respite for the larger seabirds, especially petrels, was short. When Europeans arrived, Norway and Ship Rats and cats soon became established. Later, stoats, weasels and ferrets were deliberately introduced. In a period of just a few decades these predators exterminated most of the remaining mainland petrel and shearwater colonies, so that today only a few survive. These include the Westland Petrel, one of the few petrels large and aggressive enough to see off most introduced predators, Hutton's Shearwater, which nests high in mountains where there are few mammalian predators, and Grey-faced Petrel, whose colonies still exist on some inaccessible cliff ledges.

Offshore islands became the only places where petrels and shearwaters were able to breed in any numbers, and even some of these were colonised by cats and rats. In addition a large, flightless, omnivorous native rail, the Weka (*Gallirallus australis*) was introduced, mainly by the early whalers and sealers, to various islands as an alternative food supply. Weka inflicted huge casualties on the seabirds on these islands, killing adults and chicks.

A current matter of major concern is the huge loss of pelagic birds caused by offshore fishing. It is estimated that between 100,000 and 200,000 albatrosses and smaller pelagic seabirds are lost each year to fishing boats harvesting squid and toothfish in subantarctic and Antarctic waters. Many of these birds are long-lived, slow breeders and it is quite possible that species such as the Southern Royal Albatross and the Black-browed Mollymawk could become extinct unless fishing methods less harmful to seabirds are developed and, more importantly, enforced.

On the positive side, a few seabirds have benefited from the arrival of humans in New Zealand. Gulls in particular have flourished, but in some respects this has been a double-edged sword; for example, the now abundant Black-backed Gull harasses tern and dotterel colonies, stealing eggs and chicks.

Birdwatching protocol

Good views of seabirds can usually be obtained without disturbing them. Besides, touching and handling seabirds is illegal and completely unnecessary.

Penguins, in particular, are often harassed by unthinking visitors. On no account should the birds be approached too closely. In the breeding and moulting seasons, they are especially vulnerable, and penguin deaths have resulted from people wandering through their colonies at this time.

Many seabird species no longer nest on the New Zealand mainland, because of introduced predators and the former harvesting of adults and chicks for food. Seabirds which once nested on the mainland in huge colonies now only survive in much diminished numbers on offshore islands, where they are free from disturbance. In the best interests of the seabirds, many of these islands have been closed to visitors. Information on which islands can be visited is available from the local Department of Conservation (DOC) office.

© Brian Parkinson

*The Kea (*Nestor notabilis*) is known to take chicks of Hutton's Shearwater from burrows high in the Seaward Kaikoura Range.*

Geography

This book covers the three main islands of New Zealand (North, South and Stewart Islands) together with its subantarctic and subtropical islands. The Ross Dependency in Antarctica is not included. The distribution map provided with each species in the text shows the approximate range and breeding areas within the New Zealand region. A general map of this region is given on page 12.

Topography of a seabird

Although the use of technical language is kept to a minimum, some scientific terminology, particularly as it applies to the parts of a bird, is unavoidable. The illustration shows the relevant features.

In the Guide to Species, the size of each species is given in centimetres, and is the length from bill tip to tail tip.

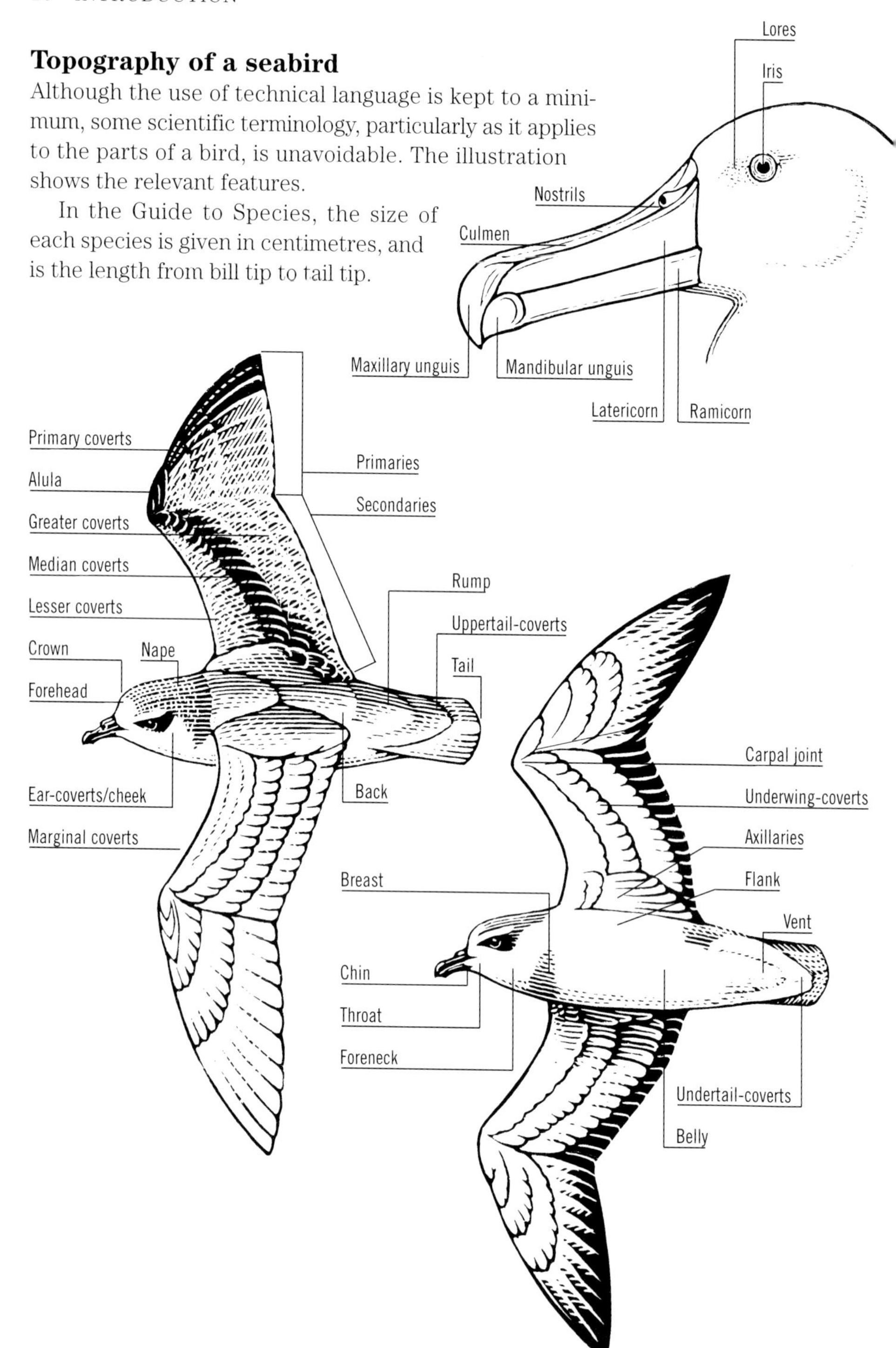

Taxonomy

In most cases taxonomy follows that of the *Checklist of the Birds of New Zealand* (Turbott, E.G., 1990, Random Century, Auckland, in association with the Ornithological Society of New Zealand Inc.). However, because significant advances have been made in recent years in seabird taxonomy, there are some changes.

The Phalacrocoracidae (shags) follow the order as set out in the *Handbook of Australian, New Zealand & Antarctic Birds* (Marchant, S. and Higgins, P. [Eds]. 1990, Oxford University Press, Melbourne). Diving petrels are elevated to family status, the Pelecanoididae, following *Seabirds: An Identification Guide* (Harrison, P. 1996, Christopher Helm Ltd, London). Major revisions have been made, and are continuing, in the Diomedeidae (albatrosses and mollymawks). The arrangement of the Diomedeidae given in this book is therefore best regarded as an interim one. It is based in large part on 'Towards a new taxonomy for albatrosses' (in Robertson, G. and Gales, R. [Eds], *Albatross Biology and Conservation*, 1998. Surrey Beatty & Sons, Chipping Norton). This in turn was based in part on the paper 'Evolutionary relationships among extant albatrosses (Procellariiformes: Diomedeidae) established from complete cytochrome-B Gene sequences', Nunn, G.B., Cooper, J., Jouventin, P., Robertson, C.J.R. and Robertson, G.G. (*The Auk*, 113, 784–801, 1996).

Order Procellariiformes: Tube-nosed birds

Family Diomedeidae: albatrosses and mollymawks
Family Procellariidae: shearwaters, petrels, fulmars, prions
Family Pelecanoididae: diving petrels
Family Oceanitidae: storm petrels

Order Sphenisciformes: Penguins

Family Spheniscidae: penguins

Order Pelecaniformes: Pelicans, gannets and shags

Family Phaethontidae: tropicbirds
Family Sulidae: gannets and boobies
Family Phalacrocoracidae: shags
Family Fregatidae: frigatebirds

Order Charadriiformes: Waders, skuas, gulls and terns

Family Stercorariidae: skuas
Family Laridae: gulls, terns and noddies

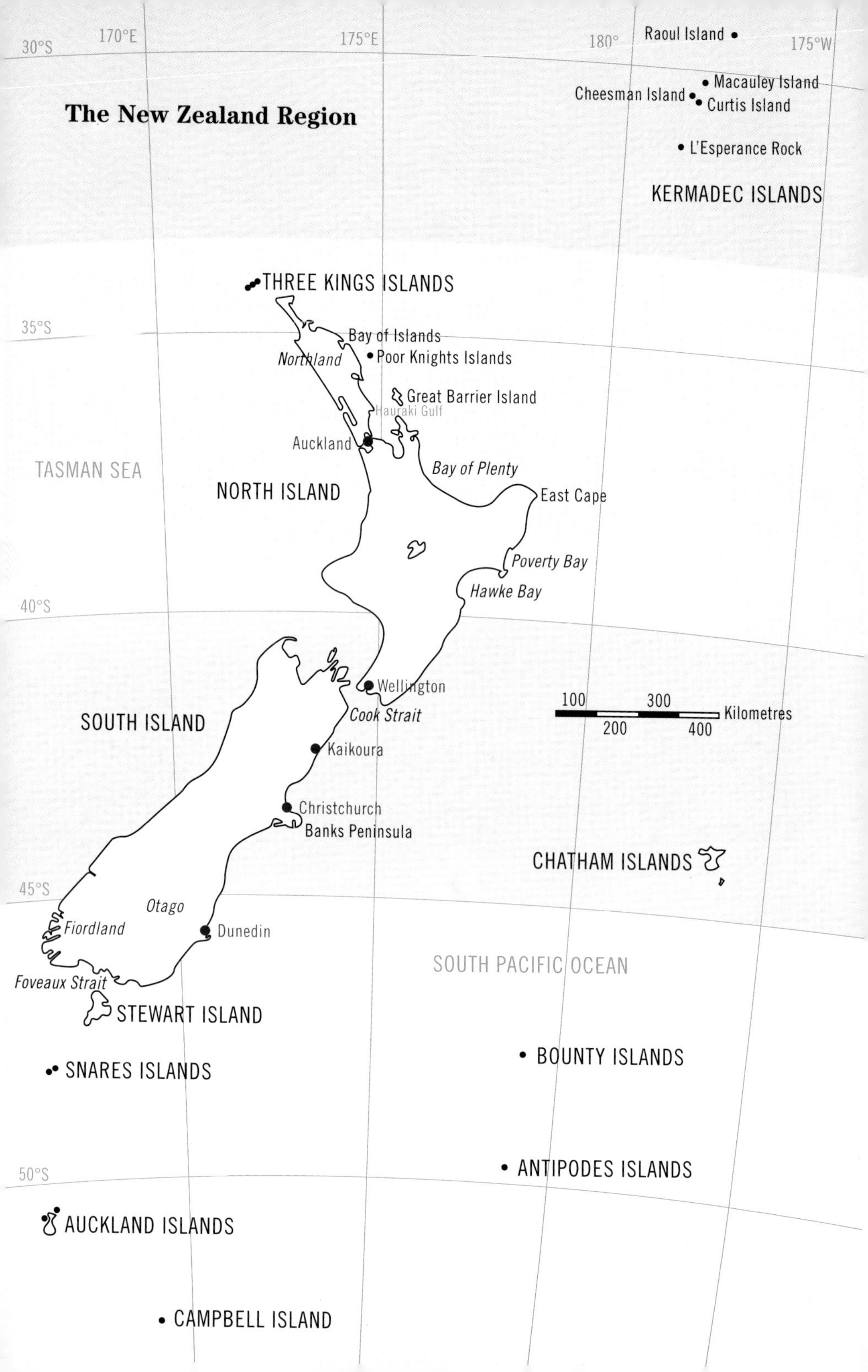

The New Zealand Region
30°S
170°E
175°E
180°
175°W
Raoul Island
Macauley Island
Cheesman Island
Curtis Island
L'Esperance Rock
KERMADEC ISLANDS
THREE KINGS ISLANDS
35°S
Bay of Islands
Northland
Poor Knights Islands
Great Barrier Island
Hauraki Gulf
Auckland
TASMAN SEA
NORTH ISLAND
Bay of Plenty
East Cape
Poverty Bay
Hawke Bay
40°S
Wellington
Cook Strait
100
300
200
400
Kilometres
SOUTH ISLAND
Kaikoura
Christchurch
Banks Peninsula
CHATHAM ISLANDS
45°S
Otago
Fiordland
Dunedin
SOUTH PACIFIC OCEAN
Foveaux Strait
STEWART ISLAND
SNARES ISLANDS
BOUNTY ISLANDS
50°S
ANTIPODES ISLANDS
AUCKLAND ISLANDS
CAMPBELL ISLAND

Checklist of species and locality guide

Which seabirds are likely to be seen varies somewhat from one end of the country to the other. The table below lists those species most likely to be seen from places where boat trips offshore are available. Popular trips include the Cream Trip (Bay of Islands), Poor Knights Islands out of Tutukaka trips, Great Barrier Island ferries, White Island excursions, Cook Strait ferries, Kaikoura Whale Watch and Seabird trips, cruises from Dunedin, and the Foveaux Strait ferries.

New Zealand breeding species are indicated on the table in the status column as (NZ); annual visitors (AV) and vagrants (V). The localities are given in four regions on the table, and correspond to the map opposite:

1 The Kermadec Islands and seas north of New Zealand.
2 Northland, Auckland, the Bay of Plenty, Poverty Bay and Hawke Bay.
3 Cook Strait, Kaikoura, the Marlborough Sounds, Westland, Banks Peninsula, and the Chatham Islands.
4 Otago, Southland, Fiordland, Stewart Island and the New Zealand subantarctic islands.

Species	Status	Region 1	Region 2	Region 3	Region 4
Snowy (Wandering) Albatross *Diomedea exulans*	AV		■	■	■
Gibson's (Wandering) Albatross *Diomedea gibsoni*	NZ	■	■	■	■
Antipodean (Wandering) Albatross *Diomedea antipodensis*	NZ	■	■	■	■
Southern Royal Albatross *Diomedea epomophora*	NZ		■	■	■
Northern Royal Albatross *Diomedea sanfordi*	NZ		■	■	■
Black-browed Mollymawk *Thalassarche melanophrys*	NZ	■	■	■	■
Campbell Mollymawk *Thalassarche impavida*	NZ	■	■	■	■
Shy (Tasmanian) Mollymawk *Thalassarche cauta*	AV/V		■	■	■
White-capped Mollymawk *Thalassarche steadi*	NZ		■	■	■

Species	Status	Region 1	Region 2	Region 3	Region 4
Salvin's Mollymawk *Thalassarche salvini*	NZ		■	■	■
Chatham Island Mollymawk *Thalassarche eremita*	NZ			■	■
Grey-headed Mollymawk *Thalassarche chrysostoma*	NZ		■	■	■
Indian Yellow-nosed Mollymawk *Thalassarche carteri*	AV		■	■	■
Buller's Mollymawk *Thalassarche bulleri*	NZ		■	■	■
Pacific Mollymawk *Thalassarche nov. sp. / platei*	NZ		■	■	■
Sooty Albatross *Phoebetria fusca*	V				■
Light-mantled Sooty Albatross *Phoebetria palpebrata*	NZ	■	■	■	■
Flesh-footed Shearwater *Puffinus carneipes*	NZ		■	■	
Wedge-tailed Shearwater *Puffinus pacificus*	NZ	■			
Buller's Shearwater *Puffinus bulleri*	NZ		■	■	■
Sooty Shearwater *Puffinus griseus*	NZ	■	■	■	■
Short-tailed Shearwater *Puffinus tenuirostris*	AV	■	■		
Fluttering Shearwater *Puffinus gavia*	NZ		■	■	
Hutton's Shearwater *Puffinus huttoni*	NZ			■	
Little Shearwater *Puffinus assimilis*	NZ	■	■	■	■
Common Diving-petrel *Pelecanoides urinatrix*	NZ		■	■	■
South Georgian Diving-petrel *Pelecanoides georgicus*	NZ				■
Grey Petrel *Procellaria cinerea*	NZ			■	■
Black Petrel *Procellaria parkinsoni*	NZ		■		

Species	Status	Region 1	Region 2	Region 3	Region 4
Westland Petrel *Procellaria westlandica*	NZ			■	■
White-chinned Petrel *Procellaria aequinoctialis*	NZ			■	■
Snares Cape Pigeon *Daption capense australe*	NZ	■	■	■	■
Southern Cape Pigeon *Daption capense capense*	AV	■	■	■	■
Antarctic Petrel *Thalassoica antarctica*	V				■
Antarctic Fulmar *Fulmarus glacialoides*	AV			■	■
Southern Giant Petrel *Macronectes giganteus*	AV	■	■	■	■
Northern Giant Petrel *Macronectes halli*	NZ	■	■	■	■
Fairy Prion *Pachyptila turtur*	NZ		■	■	■
Fulmar Prion *Pachyptila crassirostris*	NZ			■	■
Thin-billed Prion *Pachyptila belcheri*	AV			■	■
Antarctic Prion *Pachyptila desolata*	NZ		■	■	■
Salvin's Prion *Pachyptila salvini*	AV			■	■
Broad-billed Prion *Pachyptila vittata*	NZ			■	■
Blue Petrel *Halobaena caerulea*	AV			■	■
Pycroft's Petrel *Pterodroma pycrofti*	NZ		■		
Cook's Petrel *Pterodroma cookii*	NZ		■		■
Black-winged Petrel *Pterodroma nigripennis*	NZ	■	■	■	
Chatham Petrel *Pterodroma axillaris*	NZ			■	
Mottled Petrel *Pterodroma inexpectata*	NZ			■	■

Species	Status	Region 1	Region 2	Region 3	Region 4
White-naped Petrel *Pterodroma cervicalis*	NZ	■			
Kermadec Petrel *Pterodroma neglecta*	NZ	■			
Grey-faced Petrel *Pterodroma macroptera*	NZ	■	■		
Chatham Island Taiko *Pterodroma magentae*	NZ			■	
White-headed Petrel *Pterodroma lessonii*	NZ			■	■
Soft-plumaged Petrel *Pterodroma mollis*	NZ			■	■
Wilson's Storm Petrel *Oceanites oceanicus*	V	■	■	■	■
Grey-backed Storm Petrel *Oceanites nereis*	NZ		■	■	■
White-faced Storm Petrel *Pelagodroma marina*	NZ	■	■	■	■
Black-bellied Storm Petrel *Fregetta tropica*	NZ	■			■
White-bellied Storm Petrel *Fregetta grallaria*	NZ	■			
Yellow-eyed Penguin *Megadyptes antipodes*	NZ				■
Gentoo Penguin *Pygoscelis papua*	V				■
Blue Penguin *Eudyptula minor*	NZ		■	■	■
White-flippered Penguin *Eudyptula albosignata*	NZ			■	
Rockhopper Penguin *Eudyptes chrysocome*	NZ				■
Fiordland Crested Penguin *Eudyptes pachyrhynchus*	NZ				■
Snares Crested Penguin *Eudyptes robustus*	NZ				■
Erect-crested Penguin *Eudyptes sclateri*	NZ				■
Red-tailed Tropicbird *Phaethon rubricauda*	NZ	■	■		

Species	Status	Region 1	Region 2	Region 3	Region 4
White-tailed Tropicbird *Phaethon lepturus*	V	■	■		
Australasian Gannet *Morus serrator*	NZ		■	■	■
Brown Booby *Sula leucogaster*	V	■	■		
Masked Booby *Sula dactylatra*	NZ	■			
Black Shag *Phalacrocorax carbo*	NZ		■	■	■
Pied Shag *Phalacrocorax varius*	NZ		■	■	■
Little Black Shag *Phalacrocorax sulcirostris*	NZ		■	■	
Little Shag *Phalacrocorax melanoleucos*	NZ		■	■	■
King Shag *Phalacrocorax carunculatus*	NZ			■	
Stewart Island Shag *Phalacrocorax chalconotus*	NZ				■
Chatham Island Shag *Phalacrocorax onslowi*	NZ			■	
Bounty Island Shag *Phalacrocorax ranfurlyi*	NZ				■
Auckland Island Shag *Phalacrocorax colensoi*	NZ				■
Campbell Island Shag *Phalacrocorax campbelli*	NZ				■
Spotted Shag *Phalacrocorax punctatus*	NZ		■	■	■
Pitt Island Shag *Phalacrocorax featherstoni*	NZ			■	
Greater Frigatebird *Fregata minor*	V	■	■	■	
Lesser Frigatebird *Fregata ariel*	V	■	■	■	
Brown Skua *Catharacta skua*	NZ			■	■
South Polar Skua *Catharacta maccormicki*	AV			■	■

Species	Status	Region 1	Region 2	Region 3	Region 4
Arctic Skua *Stercorarius parasiticus*	AV	■	■	■	■
Pomarine Skua *Stercorarius pomarinus*	AV	■	■	■	■
Black-backed Gull *Larus dominicanus*	NZ	■	■	■	■
Red-billed Gull *Larus novaehollandiae*	NZ		■	■	■
Black-billed Gull *Larus bulleri*	NZ		■	■	■
White-winged Black Tern *Chlidonias leucopterus*	AV		■	■	■
Gull-billed Tern *Gelochelidon nilotica*	V		■	■	
Black-fronted Tern *Sterna albostriata*	NZ		■	■	■
Caspian Tern *Hydroprogne caspia*	NZ		■	■	■
White-fronted Tern *Sterna striata*	NZ		■	■	■
Sooty Tern *Sterna fuscata*	NZ	■			
Antarctic Tern *Sterna vittata*	NZ				■
Fairy Tern *Sterna nereis*	NZ		■		
Little Tern *Sterna albifrons*	AV		■	■	■
Crested Tern *Sterna bergii*	V	■	■	■	
Common Tern *Sterna hirundo*	V		■	■	
Common Noddy *Anous stolidus*	NZ	■			
White-capped Noddy *Anous tenuirostris*	NZ	■			
Grey Ternlet *Procelsterna cerulea*	NZ	■	■		
White Tern *Gygis alba*	NZ	■			

Guide to Species

Snowy (Wandering) Albatross / Toroa

Diomedea exulans

Identification 120 cm. Largest albatross. Adult white body and black upperwings, but these become progressively paler with age, so that older birds have heads, backs and inner upperwings almost entirely white. **Similar species 1.** Younger birds difficult to differentiate from other wandering albatrosses at sea, but older birds generally much paler, and bill of Snowy Albatross appears more robust at close range. **2.** Northern Royal Albatross (p. 24) has almost entirely black upperwings while Southern Royal (p. 22) has white upperwings with black primaries and some blackish barring on upperwing coverts, giving a 'dusty' appearance. **3.** Royal Albatrosses have black edge to upper mandible. In flight they may appear more 'hump-backed'. **Status** Common visitor. **Notes 1.** Previously *Diomedea exulans chionoptera*. **2.** A frequent vessel follower. **3**. Like some other albatrosses, threatened by longline fishing in the southern oceans.

Range Breeds biennially on Macquarie Island in Australian subantarctic region and on various islands in southern Indian Ocean. Ranges widely through southern oceans in non-breeding years, common visitor to New Zealand waters.
Where to see Seas around southern New Zealand, especially in winter.

Gibson's (Wandering) Albatross / Toroa

Diomedea gibsoni

Identification 115 cm. Large, with variable plumage according to age. Adult has white body and black upperwings, but younger birds can be entirely dark brown, apart from white face mask and paler underwings. **Similar species 1.** Sometimes difficult to separate from other wandering albatrosses at sea, but Gibson's always paler than Antipodean Albatross (p. 22), although not as pale as Snowy Albatross (above). **2.** Northern Royal Albatross (p. 24) has almost entirely black upperwings while Southern Royal (p. 22) has white upperwings with black primaries and some blackish barring on upperwing coverts, giving a 'dusty' appearance. **3.** Royal Albatrosses have black edge to upper mandible. In flight may appear more 'hump-backed'. **Status** Uncommon endemic. **Notes 1.** Previously *Diomedea exulans gibsoni*. **2.** A frequent vessel follower. **3.** Known to be a significant bycatch in the longline tuna fishery in New Zealand waters.

Range Breeds biennially on Auckland Islands. In non-breeding years ranges south as far as pack ice and north into subtropics. Commonly encountered on trans-Tasman shipping routes, and a frequent visitor to seas off Sydney.
Where to see Seas around New Zealand, especially in more southerly coastal waters in winter.

Snowy (Wandering) Albatross / Toroa

Gibson's (Wandering) Albatross / Toroa

Antipodean (Wandering) Albatross / Toroa

Diomedea antipodensis

Identification 115 cm. Darkest of the wandering albatrosses. Adult male has white body and black upperwings. Older males often have mottled brown patch on crown and brown smudge on chest. Females and younger birds may have entirely dark bodies. **Similar species 1.** Gibson's (p. 20) is always paler than Antipodean Albatross; Snowy Albatross (p. 20) is paler still. **2.** Northern Royal Albatross (p. 24) has almost entirely black upperwings while Southern Royal (below) has white upperwings with black primaries and some blackish barring on upperwing coverts, giving a 'dusty' appearance. **3.** Royal Albatrosses have black edge to upper mandible. In flight may appear more 'hump-backed'. **Status** Uncommon endemic. **Notes 1.** Previously *Diomedea exulans antipodensis*. **2.** A frequent vessel follower. **3.** Known to be a relatively frequent bycatch in the New Zealand longline fishery.

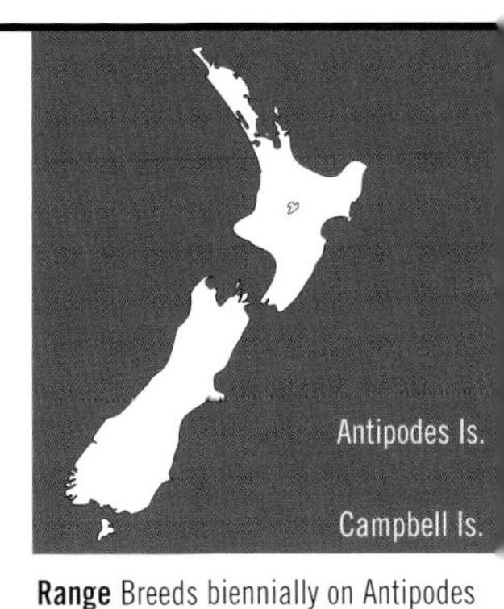

Range Breeds biennially on Antipodes Islands with a few pairs on Campbell Island. In non-breeding years ranges widely through South Pacific.
Where to see Seas around New Zealand, especially in more southerly coastal waters in winter.

Southern Royal Albatross / Toroa-whakaingo

Diomedea epomophora

Identification 115 cm. Body white with white upperwings which have black primaries and some blackish barrings on the primary coverts, giving a 'dusty' appearance. Juveniles some black flecking on upperparts. Bill flesh-coloured with a diagnostic black cutting edge to the upper mandible. Feet pinkish. **Similar species 1.** Northern Royal (p. 24) has almost entirely black upperwings while Southern Royal has much whiter upperwings. **2.** Southern Royal distinguished from wandering albatrosses at close range by black edge to the upper mandible. Also Royals have different upperwings and lack brown body mottling of younger wandering albatrosses. **Status** Locally common endemic. **Notes 1.** Previously *Diomedea epomophora epomophora*. **2.** Threatened by longline fishing in the southern oceans. **3.** Biennial breeder, like Northern Royal.

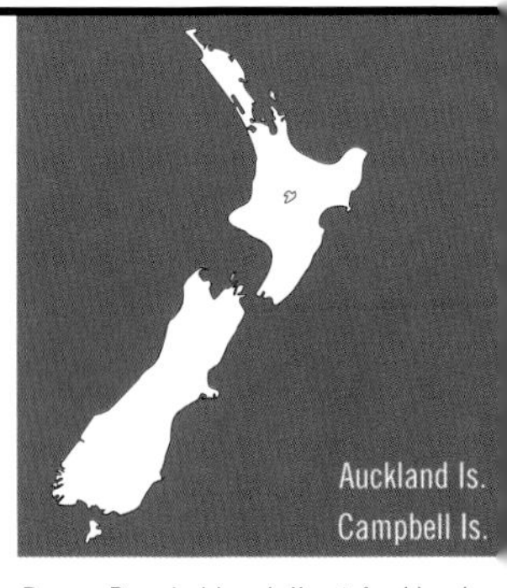

Range Breeds biennially at Auckland and Campbell Islands. May circumnavigate southern hemisphere.
Where to see Off southern coasts of New Zealand in winter and around subantarctic breeding sites.

Antipodean (Wandering) Albatross / Toroa

Southern Royal Albatross / Toroa-whakaingo

Northern Royal Albatross / Toroa-whakaingo

Diomedea sanfordi

Identification 115 cm. Body and back white with black upperwings. Juveniles resemble adults, but may have some dark flecking on crown and back. Feet pinkish and bill flesh-coloured with diagnostic black cutting edge to upper mandible. **Similar species 1.** In Northern Royal, upperwings are almost entirely black, while Southern Royal (p. 22) appears much whiter overall, with white upperwings which have black primaries and some blackish barring on upperwing coverts, giving a 'dusty' appearance. **2.** Northern Royal distinguished from wandering albatrosses at close range by black edge to upper mandible. Also Royals have different upperwings and lack brown body mottling of younger wandering albatrosses. **Status** Locally common endemic. **Notes 1.** Previously *Diomedea epomophora sanfordi*. **2.** Climatic changes and erosion at its Chatham Island breeding sites are causing lower breeding success.

Range Breeds biennially at Taiaroa Head near Dunedin and at Chatham Islands. Regularly ranges to northern New Zealand waters.
Where to see Taiaroa Head, off the New Zealand coast and around the Chatham Islands.

Black-browed Mollymawk / Toroa

Thalassarche melanophrys

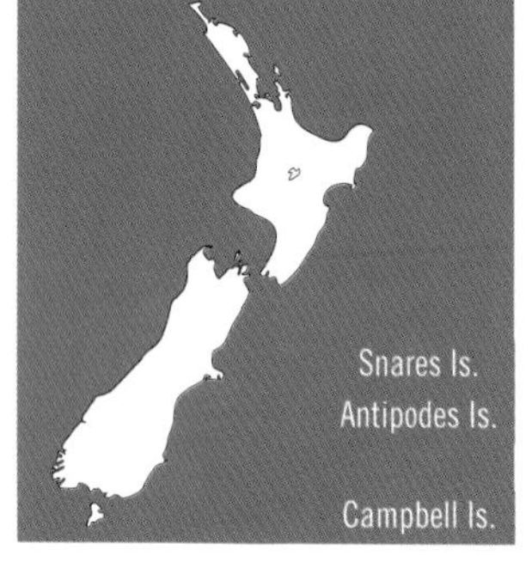

Identification 90 cm. Medium-sized with white body and head, black upperwings and white underwings with diagnostic broad, black margins. Bill yellowish-orange with reddish tip. Legs and feet pale bluish-white. Prominent black eyebrows and dark iris give distinctive scowling appearance. Immatures have dusky bills, some grey on crown and neck, forming a collar, and dusky centres to underwings. **Similar species** Campbell Mollymawk (p. 26), but black brow of the Black-browed less obvious, iris is dark and black underwing margins not quite as broad. **Status** Common native. **Notes 1.** Previously *Diomedea melanophrys melanophrys*. **2.** Probably most commonly seen mollymawk in northern New Zealand waters. **3.** Keen follower of vessels.

Range Breeds annually on many islands throughout subantarctic region. In New Zealand, breeds on Campbell, Antipodes and Snares Islands. In winter, ranges widely through South Pacific as far north as tropics.
Where to see Common in coastal waters around New Zealand in winter.

Northern Royal Albatross / Toroa-whakaingo

Black-browed Mollymawk / Toroa

Campbell Mollymawk / New Zealand Black-browed Mollymawk / Toroa

Thalassarche impavida

Identification 90 cm. Medium-sized with white body and head, black upperwings and white underwings with diagnostic broad, black margins. Bill of adult yellowish-orange with reddish tip. Legs and feet pale bluish-white. Obvious black eyebrows and, at close range, honey-coloured iris visible. Immature has darker bill, some grey on crown and hind neck and very dark centres to underwings. **Similar species** Black-browed Mollymawk (p. 24), but Campbell Mollymawk has more obvious black eyebrows and wider underwing margins. In Campbell Mollymawk, iris is honey-coloured rather than black as in Black-browed Mollymawk. **Status** Common endemic. **Notes 1.** Previously *Diomedea melanophrys impavida*. **2.** Much rarer on a world scale than Black-browed Mollymawk. **3.** Keen follower of vessels. **4.** Recorded as bycatch in tuna fisheries in New Zealand and Australia.

Range Breeds annually only on Campbell Island. In winter, ranges widely throughout New Zealand coastal waters. Away from New Zealand, ranges to southern Australian waters, Tasman Sea and tropical waters of south-west Pacific.

Where to see Present in coastal waters in winter, often reaching northern New Zealand.

Shy (Tasmanian) Mollymawk

Thalassarche cauta

Identification 90 cm. Along with White-capped (p. 28), the largest mollymawk. Chest, belly and rump white, underwings white with narrow black margin. Distinctive black 'thumb mark' where anterior underwing meets body. Faint greyish smudge on cheek, bill pale yellowish-grey with yellow tip. **Similar species** Similar to White-capped Mollymawk (p. 28), but has a paler bill. **Status** Uncommon visitor. **Notes 1.** Previously *Diomedea cauta cauta*. **2.** Keen vessel follower, which may congregate around fishing boats in large flocks. **3.** Its foraging range in southern Australia overlaps with local longline fisheries where this species forms a frequent bycatch.

Range Breeds annually on three small islands off the Tasmanian coast. An occasional visitor to New Zealand waters.

Where to see Coastal waters of North, South and Stewart Islands.

© Don Hadden

Campbell Mollymawk /
New Zealand Black-browed Mollymawk / Toroa

© Peter Lansley

Shy (Tasmanian) Mollymawk

White-capped Mollymawk

Thalassarche steadi

Identification 90 cm. A large mollymawk, chest, belly and rump white, underwings white with narrow black margin. Distinctive black 'thumb mark' where anterior underwing meets body. Faint greyish smudge on cheek, bill bluish-grey with yellow tip and pale yellow culmen. **Similar species 1.** Shy Mollymawk (p. 26), but in White-capped bill is bluish-grey rather than yellowish-grey. **2.** Salvin's Mollymawk (below) but Salvin's has greyish head and neck with white forehead. Bill in Salvin's greyer with dark unguis. **Status** Locally common endemic. **Notes 1.** Previously *Diomedea cauta steadi*. **2.** Keen follower of vessels. **3.** May be at risk as bycatch in tuna and squid-trawl fisheries.

Range Breeds annually at Auckland and Antipodes Islands. Frequent visitor to coastal waters around New Zealand. **Where to see** Coastal waters from Northland (especially in winter and spring) south to Foveaux Strait. More numerous in the south.

Salvin's Mollymawk

Thalassarche salvini

Identification 85 cm. Medium-sized to large mollymawk. Chest, belly and rump white, underwings white with narrow black margin. Distinctive black 'thumb mark' where anterior underwing meets body. Head greyish-brown to brown with contrasting white forehead. Bill bluish-grey with yellowish culmen and base. Tip of upper mandible yellow while lower mandible has a diagnostic black tip (or unguis). **Similar species** Chatham Island Mollymawk (p. 30) has a much darker head and yellower bill. **Status** Locally common endemic. **Notes 1.** Previously *Diomedea cauta salvini*. **2.** Keen vessel follower. **3.** Not known to suffer significant mortality as a fisheries bycatch.

Range Breeds annually mainly at Bounty and Snares Islands (Western Chain). A frequent visitor to coastal waters around New Zealand and ranges widely in the southern oceans as far as South America and southern Africa. **Where to see** Coastal waters from Northland (especially in winter and spring) south to Foveaux Strait. More numerous in the south.

© Dennis Buurman

White-capped Mollymawk

© Don Hadden

Salvin's Mollymawk

Chatham Island Mollymawk

Thalassarche eremita

Identification 85 cm. Medium-sized to large. Chest, belly and rump white, underwings white with narrow black margin. Distinctive black 'thumb mark' where anterior underwing meets body. Head dark grey, contrasting sharply with white underparts. Bill bright yellow with distinctive dark spot on the unguis. **Similar species** Salvin's Mollymawk (p. 28), but Chatham Island Mollymawk has noticeably darker head and much yellower bill. **Status** Rare endemic. **Notes 1.** Previously *Diomedea cauta eremita*. **2.** Tends not to follow vessels. **3.** No evidence of mortality associated with fisheries bycatch.

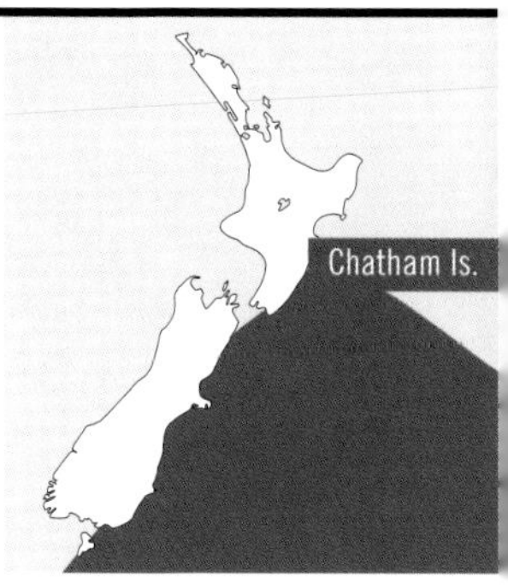

Range Breeds annually only on Pyramid Rock, Chatham Islands. Locally common around the southern Chatham Islands. Ranges to the seas east of the South Island and to the eastern South Pacific Ocean off Chile.
Where to see Coastal waters of the Chatham Islands, seas off eastern side of the South Island.

Grey-headed Mollymawk

Thalassarche chrysostoma

Identification 80 cm. Medium-sized mollymawk. Grey head with paler forehead, and glossy black bill with bright yellow culmen and base. Body white and upperwings black. Underwings white with broad black margins which are much wider on leading edge. **Similar species** Buller's Mollymawk (p. 32), but head of Buller's not so dark and has paler crown. Buller's also has slightly narrower underwing margins, especially on leading edge. **Status** Locally common native. **Notes 1.** Previously *Diomedea chrysostoma*. **2.** Does not normally follow vessels but may gather around stationary fishing boats. **3.** Threatened by longline fisheries.

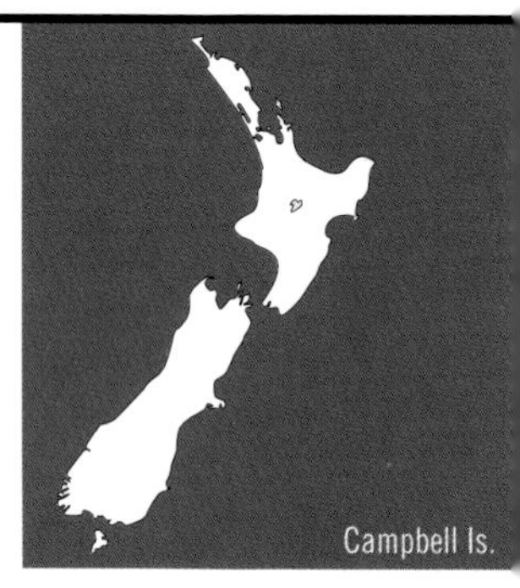

Range Widespread circumpolar breeding distribution. In New Zealand region breeds biennially only on Campbell Island. Regular visitor to South Island and Stewart Island waters.
Where to see Waters off South and Stewart Islands, especially in winter.

Chatham Island Mollymawk

Grey-headed Mollymawk

Indian Yellow-nosed Mollymawk

Thalassarche carteri

Identification 75 cm. The smallest mollymawk. Distinguished by slender silhouette and relatively narrow dark underwing margins. Head white. Bill, from a distance, appears black and slender, but at close range yellow culmen and tip are distinctive. **Similar species 1.** Buller's Mollymawk (below), from which it can be distinguished by its smaller size, whitish head and narrower underwing margins. **2.** Atlantic Yellow-nosed Mollymawk, *Thalassarche chlororhynchos*, an occasional visitor, which has grey head and white forehead patch. **Status** Uncommon visitor. **Notes 1.** May associate with Australasian Gannets (p. 92). **2.** Forms a significant bycatch in the southern bluefin tuna fishery.

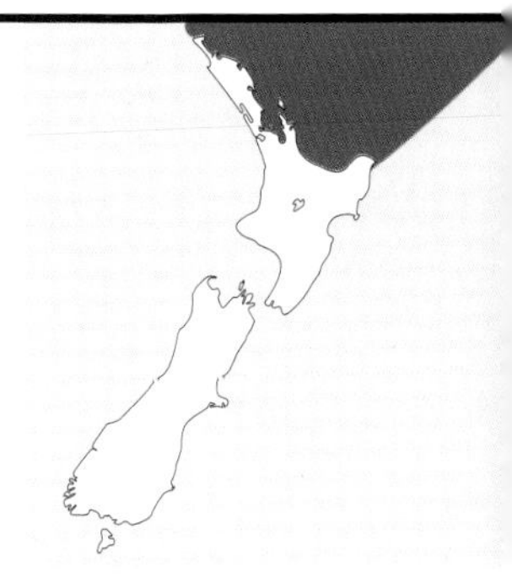

Range Breeds annually on four island groups in Indian Ocean. A regular visitor to Northland's east coast, Hauraki Gulf and Bay of Plenty waters. **Where to see** In waters of Northland's east coast, Hauraki Gulf and Bay of Plenty in winter. May occur very close inshore.

Buller's Mollymawk / Southern Buller's Mollymawk / Toroa-teoteo

Thalassarche bulleri

Identification 80 cm. Medium-sized. Body white, upperwings black, underwings white with the leading margin broader. Head grey with silvery-white patch on forehead. Black bill with prominent yellow culmen and base. Legs and feet bluish. **Similar species** Closely resembles Pacific Mollymawk (p. 34), but in Buller's forehead is silvery-white while in Pacific forehead is silver-grey and the latter has a more robust bill. Difficult to distinguish from Pacific at sea. **Status** Locally common endemic. **Notes 1.** Previously *Diomedea bulleri bulleri*. **2.** An avid follower of vessels and voracious consumer of fish offal. **3.** Heavy mortality reported in trawl and longline fisheries.

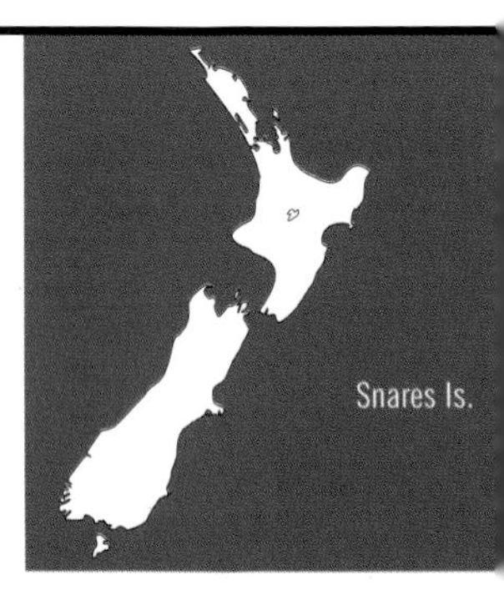

Range Breeds annually at Snares Islands and the Solanders. Ranges through more southerly coastal waters, dispersing to the eastern Pacific in the non-breeding season. **Where to see** Coastal waters of the South and Stewart Islands.

Indian Yellow-nosed Mollymawk

Buller's Mollymawk / Southern Buller's Mollymawk / Toroa-teoteo

Pacific Mollymawk / Northern Buller's Mollymawk / Toroa-teoteo

Thalassarche nov. sp. / platei

Identification 80 cm. Medium-sized. Body white; upperwings black, underwings white with black margins with the leading margins broader. Head grey with silver-grey patch on forehead. Black bill with prominent yellow culmen and base. Legs and feet bluish. Immatures of both Buller's and Pacific Mollymawks have entirely dark grey heads with dusty brown, dark-tipped bills. **Similar species** Closely resembles Buller's Mollymawk (p. 32), but in the Pacific head is grey with a silver-grey patch on forehead, while in Buller's forehead silvery-white. Bill in Pacific is more robust. Difficult to distinguish from Buller's at sea. **Status** Locally common endemic. **Notes 1.** Previously *Diomedea bulleri platei*. **2.** A keen vessel follower. **3.** No evidence it is threatened as bycatch in fisheries.

Range Breeds annually on The Sisters and Forty-four Islands off the Chatham Islands and at Rosemary Rocks off the Three Kings Islands. Ranges throughout coastal New Zealand waters, dispersing to the eastern Pacific in non-breeding season.
Where to see Coastal waters of the Chatham Islands, South Island and around the Three Kings Islands in summer.

Sooty Albatross

Phoebetria fusca

Identification 85 cm. Entirely sooty brown body becoming slightly darker on head and wings, which are long and narrow. Tail wedge-shaped. White eye ring obvious at close quarters. Bill black and slender with yellow line along the lower mandible. **Similar species** At a distance could be confused with Light-mantled Sooty Albatross (p. 36), but Light-mantled has a much lighter body. **Status** Rare visitor. **Notes** Sometimes trails vessels for long periods.

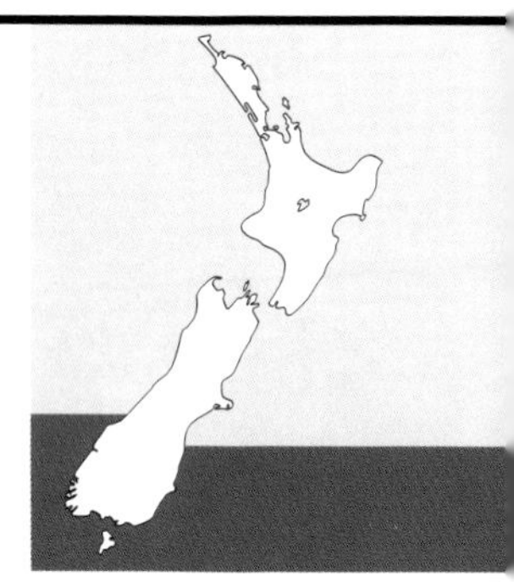

Range Breeds biennially on subantarctic islands in the Indian and Atlantic Oceans. Ranges widely in the subantarctic region and occasionally reaches New Zealand subantarctic waters.
Where to see Waters of the subantarctic region, but rarely north of the 40th parallel (north of Marlborough Sounds, approximately).

Pacific Mollymawk / Northern Buller's Mollymawk / Toroa-teoteo

Sooty Albatross

Light-mantled Sooty Albatross / Toroa-haunui

Phoebetria palpebrata

Identification 80 cm. Grey body with contrasting sooty-brown head and wings. Wings and tail long and pointed. White eye ring obvious at close range. Bill black and slender with faint yellow line along lower mandible. **Similar species** At a distance could be confused with Sooty Albatross (p. 34), but Light-mantled has much paler body. **Status** Uncommon native. **Notes 1.** Sometimes appears near the coast in periods of high winds. (The Maori name 'haunui' means 'big wind'.) **2.** Follows vessels for long periods.

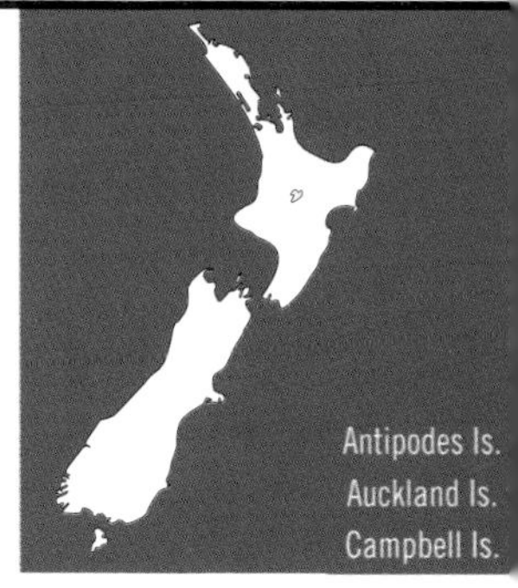

Range Breeds widely on subantarctic islands. In New Zealand region, breeds on Auckland, Antipodes and Campbell Islands. Ranges widely in New Zealand subantarctic waters, occasionally reaching southern coastal waters. Some may range into seas north of New Zealand.
Where to see Waters off and to south of Stewart Island. Occasionally ranges north of the 40th parallel.

© Peter Langlands

Light-mantled Sooty Albatross / Toroa-haunui

Flesh-footed Shearwater / Toanui

Puffinus carneipes

Identification 44 cm. A large, all-dark shearwater with pale pinkish bill with a dark tip. Distinctive pink feet. Slow-flapping and buoyant flight. **Similar species 1.** Sooty Shearwater (p. 40), but lacks the Sooty's pale underwings, and its flight pattern is different: the Sooty's flight consists of rapid wingbeats interspersed with long glides. **2.** Black Petrel (p. 48), but this is blacker with ivory-coloured bill, and black feet which just project beyond the tail. **3.** Grey-faced Petrel (p. 72), but this has stout black bill, and powerful stiff-winged flight. **Status** Common native. **Notes** Often attracted to stationary vessels, where it scavenges bait from fishing lines.

Range Breeds on a number of islands off the north-east coast of the North Island, the Auckland west coast, Taranaki and Cook Strait. Also breeds in large numbers on Lord Howe Island and on some Indian Ocean Islands. Migrates to the North Pacific during the southern winter.
Where to see In coastal waters off the Northland east coast, Hauraki Gulf and Bay of Plenty in summer.

Wedge-tailed Shearwater / Koakoa

Puffinus pacificus

Identification 46 cm. Large, all-dark, slightly built shearwater with prominent wedge-shaped tail. Grey bill and flesh-coloured feet. Buoyant, leisurely flight with wings held well forward with much shearwatering. **Similar species** Sooty Shearwater (p. 40), but the Wedge-tailed has dark underwings and long, wedge-shaped tail. The ranges of the two species do not usually overlap, except perhaps when the two species are on migration. **Status** Locally common native. **Notes 1.** The race breeding at the Kermadec Islands is *P. p. pacificus*. **2.** Not generally a vessel follower but may accompany pods of dolphins.

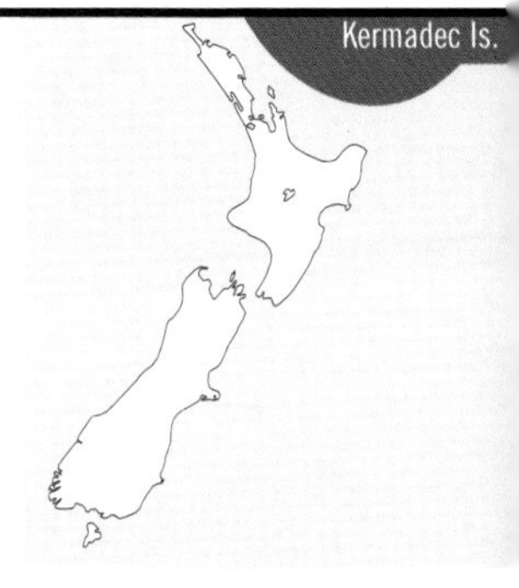

Range Breeds on Kermadec Islands, on islands off Australian coast and on many islands in tropical and subtropical zones of Indian and Pacific Oceans.
Where to see Unlikely to be seen in coastal New Zealand waters. Best place is around the Kermadec Islands during summer.

Flesh-footed Shearwater / Toanui

Wedge-tailed Shearwater / Koakoa

Buller's Shearwater / Rako

Puffinus bulleri

Identification 46 cm. Large, slender-bodied, long-tailed, greyish-brown shearwater with distinctive 'M' pattern on upperwings, and white underparts. Has graceful, gliding flight interspersed with strong wingbeats. **Similar species** Distinctive upperwing pattern distinguishes Buller's from other shearwaters. **Status** Common endemic. **Notes 1.** Formerly taken as muttonbirds (young birds taken from nesting burrows) by northern Maori. **2.** Almost wiped out by feral pigs before 1936 on Aorangi Island. Pigs were removed and the population has recovered. **3.** Not to be confused with Bulwer's Petrel, *Bulweria bulwerii*, which is a much smaller, all-dark species, not found in New Zealand waters.

Range Breeds in summer in large numbers only on the Poor Knights Islands. Common summer resident of coastal New Zealand waters, especially in the north. Migrates to the seas off the western coasts of North America during the southern winter.
Where to see Seas off Northland, Hauraki Gulf and coastal Bay of Plenty.

Sooty Shearwater / Titi

Puffinus griseus

Identification 44 cm. A large, all-dark shearwater usually with a pale silvery underwing. Bill and feet dark. Distinctive fast-flapping flight with long glides. **Similar species** Short-tailed Shearwater (p. 42), but when the two species are seen together, the Sooty is somewhat larger with lighter underwings and a longer bill. **Status** Abundant native. **Notes** The Sooty is the most commonly harvested muttonbird in New Zealand, the young being taken from nesting burrows in their hundreds of thousands in May by southern Maori.

Range Abundant summer resident, often in large flocks off the South and Stewart Islands. The biggest breeding colonies exist around Stewart, Snares and Chatham Islands. Also occurs off Chile and in the South Atlantic. New Zealand birds migrate to the North Pacific during the southern winter.
Where to see In coastal waters around New Zealand, especially in the south.

Buller's Shearwater / Rako

Sooty Shearwater / Titi

Short-tailed Shearwater

Puffinus tenuirostris

Identification 42 cm. A medium-sized to large, all-dark shearwater, with dusky grey underwings and a short tail. Bill and feet dark. Its flight is a similar fast-flap and glide sequence to that of Sooty Shearwater but more exuberant, with much wheeling and plunging in rough weather. **Similar species** Sooty Shearwater (p. 40), but when seen together, Short-tailed is somewhat smaller with darker underwings and shorter bill. **Status** Locally common migrant. **Notes 1.** Young birds are harvested on the Bass Strait Islands of Australia in very large numbers. **2.** Many also die in drift nets in the North Pacific fisheries.

Range Breeds in large numbers in Bass Strait (between the Australian mainland and Tasmania). Migrates to the North Pacific in the southern winter.

Where to see Not uncommon as a beach-wreck around the New Zealand coast. Many pass through New Zealand waters on their spring and autumn migrations.

Fluttering Shearwater / Pakaha

Puffinus gavia

Identification 33 cm. A medium-sized shearwater with dark brown upperparts, dark face and white underparts. Low, rapid, direct flight near the surface of the water. **Similar species 1.** Hutton's Shearwater (p. 44), but Hutton's is slightly bigger with dark patch on inner underwings. **2.** Little Shearwater (p. 44), but Little is smaller and darker with a distinctive white face. **Status** Abundant endemic. **Notes 1.** Formerly taken as muttonbirds (young birds taken from nesting burrows), but now protected. **2.** Often forms large rafts in coastal waters in calm water.

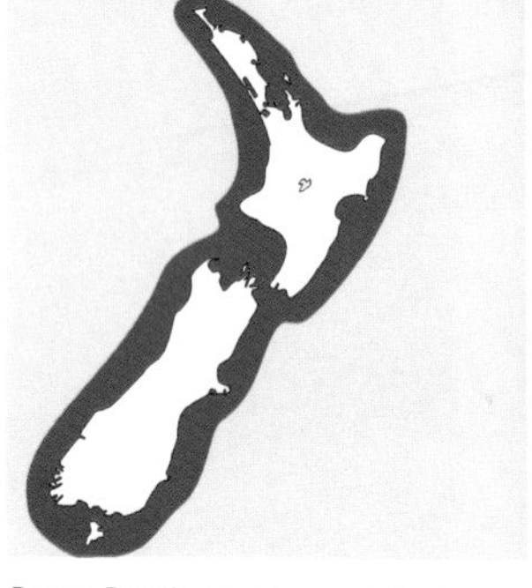

Range Breeds on many predator-free islands off the north-east coast of the North Island and islands in the Marlborough Sounds. Generally fairly sedentary, however in winter may range further offshore, sometimes as far as the Chatham Islands, with many young migrating to seas off eastern Australia.

Where to see Coastal waters of North, South and Stewart Islands.

Short-tailed Shearwater

Fluttering Shearwater / Pakaha

Hutton's Shearwater / Titi

Puffinus huttoni

Identification 36 cm. Medium-sized shearwater with dark brownish-black upperparts, dark face, brown mottling on sides of neck and chest, and white underparts. A darkish patch on inner underwing. Low, rapid, direct flight near the surface of the water. **Similar species** Fluttering Shearwater (p. 42), but Fluttering is smaller with paler underwing and lacks brown neck mottling of Hutton's. **Status** Locally common endemic.

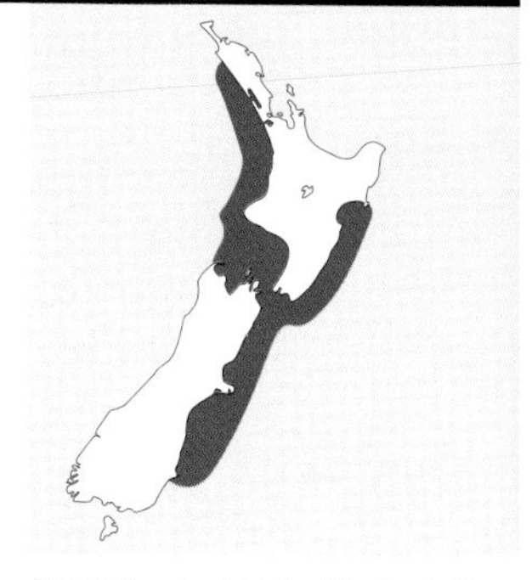

Range Breeds at high altitudes in the Seaward Kaikoura Range. Ranges along the Kaikoura Coast to Cook Strait and Tasman Sea. In winter migrates to seas around Australia.
Where to see Kaikoura whale-watching trips and from Cook Strait ferries. May also be seen when feeding in flocks close inshore off the Kaikoura and North Canterbury coasts and Banks Peninsula in summer.

Little Shearwater / Pakaha

Puffinus assimilis

Identification 30 cm. A small shearwater with black upperparts, white underparts and a distinctive white face. Flight stiff-winged, direct and rapid. **Similar species 1.** Fluttering Shearwater (p. 42), but the Little is smaller, blacker and has a white face. **2.** Common Diving Petrel (p. 46), but Little Shearwater is larger and longer-winged, and does not dive as frequently. **Status** Locally common native. **Notes 1.** At places such as the Antipodes and Chatham Islands Little Shearwaters fall prey to Brown Skuas (p. 110). **2.** Sensitive to introduced rodents, with the most successful breeding colonies confined to rodent-free islands. **3.** Three subspecies in New Zealand: Kermadec Little, *P. a. kermadecensis*; North Island Little, *P. a. haurakiensis*; Subantarctic Little, *P. a. elegans*. They are indistinguishable at sea.

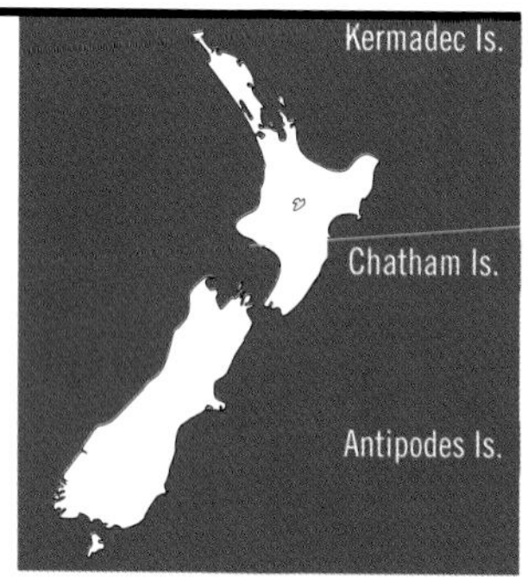

Range Kermadec Little Shearwater breeds on the Kermadec Islands. North Island Little Shearwater breeds on islands off the east coast of the North Island from the Bay of Islands south to East Cape. Subantarctic Little Shearwater breeds around the Chatham Islands and the Antipodes Islands.
Where to see Outer, more oceanic coastal waters, especially from North Cape south to around East Cape.

Hutton's Shearwater / Titi

Little Shearwater / Pakaha

Common Diving-petrel / Kuaka

Pelecanoides urinatrix

Identification 20 cm. Small, dumpy petrel, dark above, undersides dusky white. Black bill, blue legs and feet. Flight rapid and direct. Dives often. **Similar species** Difficult to distinguish from South Georgian Diving-petrel (below) at sea, but South Georgian has whiter underparts. **Status** Abundant native. **Notes 1.** Vulnerable to introduced rats and cats, and many colonies on offshore islands have been exterminated by predators. **2.** Not a strong flier; the most aquatic of all petrels, dives frequently and 'flies' penguin-like under water.

Range Circumpolar distribution around subantarctic. In New Zealand region breeds on many islands around the North Island, in Cook Strait, Foveaux Strait and Stewart Island, Chatham, Snares, Antipodes and Auckland Islands.
Where to see Coastal waters of North, South, Stewart and Chatham Islands.

South Georgian Diving-petrel

Pelecanoides georgicus

Identification 18 cm. Small, dumpy petrel, dark above and white below with short wings and rapid, direct flight. Black bill, blue legs and feet. **Similar species** Difficult to distinguish from Common Diving-petrel (above) at sea, but Common Diving-petrel has dusky white underparts. **Status** Rare native. **Notes** Only extant colony in New Zealand is in sand dunes on Codfish Island (off Stewart Island). It has shown some signs of recovery there since Weka were removed.

Range Widespread circumpolar subantarctic distribution. In New Zealand region breeds only on Codfish Island (off Stewart Island).
Where to see Possibly in waters off the west coast of Stewart Island.

Common Diving-petrel / Kuaka

South Georgian Diving-petrel

Grey Petrel / Kuia

Procellaria cinerea

Identification 48 cm. A robust, ashy-grey petrel with darkish crown and upperparts. Underparts are white, tail grey and wedge-shaped, underwings dark grey. The bill greenish-yellow, feet pinkish-grey. **Similar species** Larger grey and white shearwaters, especially Buller's (p. 40), from which Grey Petrel can be distinguished by its dark underwings and paler bill. **Status** Uncommon native. **Notes** Once common on Campbell Island, but now almost completely eradicated there by rats and feral cats.

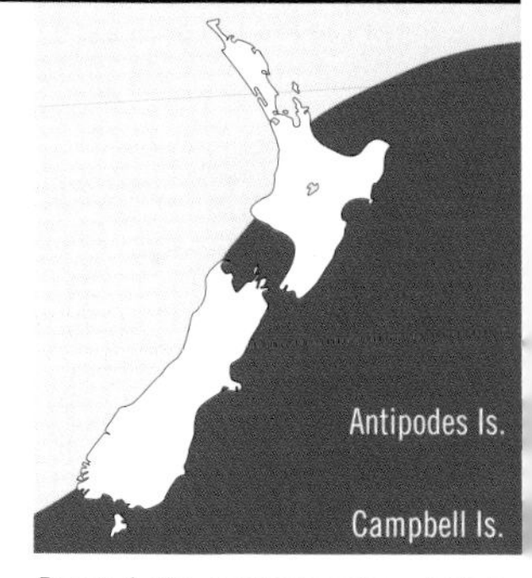

Range A circumpolar species, which in New Zealand region breeds in large numbers on the Antipodes Islands. May range as far north as the Bay of Plenty and east coast of the North Island in winter and spring.
Where to see Coastal waters of South and Stewart Islands and southern North Island in winter.

Black Petrel / Taiko

Procellaria parkinsoni

Identification 46 cm. A large, entirely brownish-black petrel with black feet. Pale yellow bill has dark tip. The dark feet extend beyond centre of tail. **Similar species** **1.** Westland Petrel (p. 50), which is larger. **2.** Flesh-footed Shearwater (p. 38), which is browner, has a pinkish bill and flesh-coloured feet that do not extend beyond the tail. **3.** Grey-faced Petrel (p. 72), which is paler, has a pale face and stout black bill. Black Petrel ranges more to the north and north-east of the North Island than these other species. **Status** Rare endemic. **Notes 1.** Eradicated by cats and other mammalian predators from many of its former mainland breeding colonies. The removal of feral cats from Little Barrier Island has been a positive step towards halting its decline. **2.** May benefit from reintroduction programmes to former breeding sites.

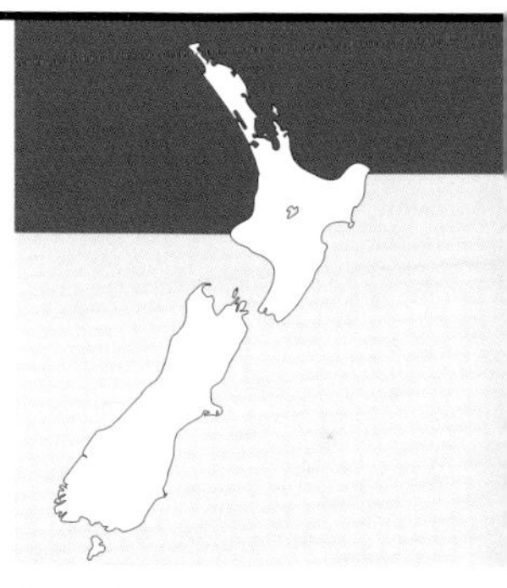

Range Breeds only on Little and Great Barrier Islands. In winter migrates to the eastern tropical Pacific Ocean.
Where to see The Hauraki Gulf, especially between Great and Little Barrier Islands and the mainland, and the seas east of Great Barrier. Ashore at breeding colony on Mt Hobson (Hirakimata) on Great Barrier Island.

© Don Hadden

Grey Petrel / Kuia

© DOC/C.R. Veitch

Black Petrel / Taiko

Westland Petrel

Procellaria westlandica

Identification 48 cm. A very large, entirely blackish-brown petrel. Bill ivory-yellow with a black tip. Legs and feet black. **Similar species 1.** Almost indistinguishable from White-chinned Petrel (below) at sea, however in hand Westland lacks white chin and has dark-tipped bill. **2.** Black Petrel (p. 48), which is smaller. The Black has a more northerly range than Westland Petrel. **Status** Uncommon endemic. **Notes 1.** Young were formerly taken from nesting burrows by Maori and early European settlers. **2.** Breeds during winter.

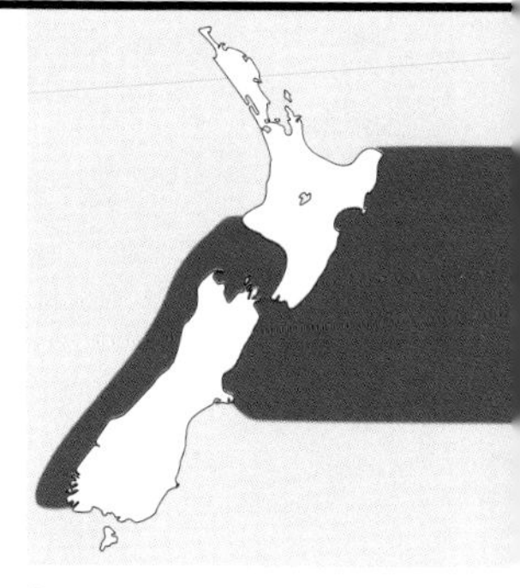

Range Breeds only in the forested coastal hills of central Westland. During the breeding season it often feeds off the Kaikoura and Westland coasts. In the non-breeding season it ranges to eastern Australia and as far afield as Peru and Chile.
Where to see In coastal waters off the west coast of the South Island, Cook Strait and Kaikoura.

White-chinned Petrel

Procellaria aequinoctialis

Identification 55 cm. A very large, robust petrel. Entirely blackish-brown apart from small white patch on the chin, although this feature not always evident. Bill pale ivory-yellow, legs and feet black. **Similar species 1.** Almost indistinguishable from Westland Petrel (above), however in hand white chin is visible and Westland Petrel has a dark-tipped bill. White-chinned Petrels tend to have a more southerly range than Westland Petrels. **2.** Black Petrel (p. 48), but Black is smaller and lacks white chin. **3.** Flesh-footed Shearwater (p. 38), but this species is smaller and browner with flesh-coloured feet and a pinkish bill. Also, in all Procellarias the feet project beyond the centre of the short, rounded tail. **Status** Uncommon native.

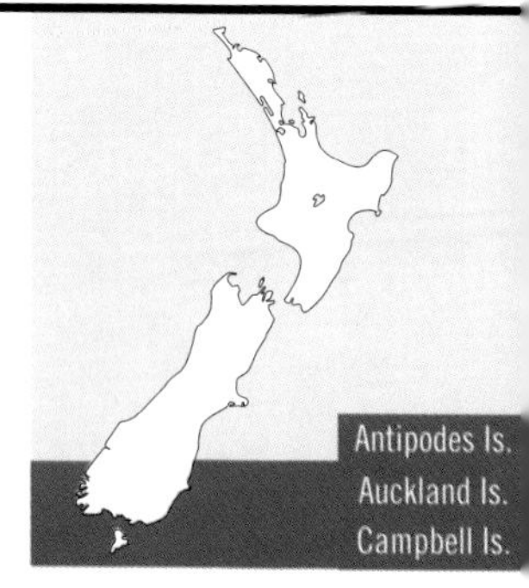

Range A circumpolar species which breeds on Auckland, Campbell, Antipodes and many other subantarctic islands. A winter visitor to Foveaux Strait and waters off Stewart Island.
Where to see Foveaux Strait and waters off Stewart Island.

Westland Petrel

White-chinned Petrel

Cape Pigeon / Titore

Daption capense

Identification 40 cm. Pigeon-sized petrel with distinctive black and white chequered upperparts, and white underparts and underwings. Two subspecies in New Zealand with differing amounts of black on upperwings: **1.** Snares Cape Pigeon, *D. c. australe*, has darker upperwings with white patches separated by black areas. **2.** Southern Cape Pigeon, *D. c. capense*, has much whiter upperwings flecked with black. **Similar species** Antarctic Petrel (below), from which Cape Pigeon can readily be distinguished by chequerboard upperwing pattern. **Status** Common native. **Notes 1.** A keen vessel follower. **2.** Formerly numerous around whaling ships and stations.

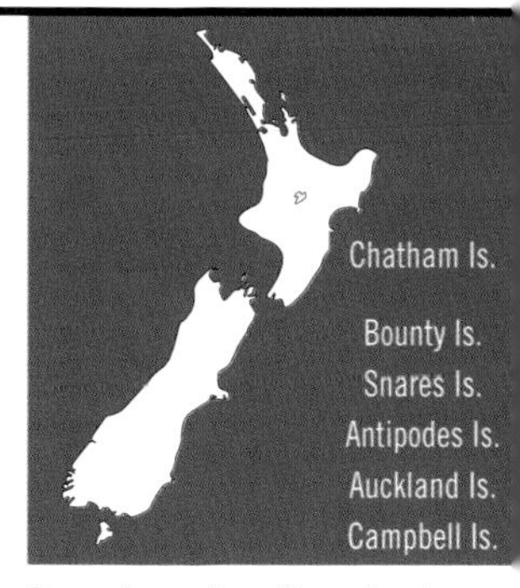

Range Snares Cape Pigeon breeds on most of outlying southern New Zealand islands, while the Southern Cape Pigeon breeds on and around the Antarctic coastline.
Where to see A regular winter visitor to the Hauraki Gulf and seas to the north-east of New Zealand. Common in coastal waters of South and Stewart Islands, and easily seen from the Cook Strait ferries.

Antarctic Petrel

Thalassoica antarctica

Identification 45 cm. The upperparts and upperwings are an uninterrupted ashy brown apart from a broad white wingbar. Tail white with dark brown tip. Bill is greyish-black. **Similar species** Cape Pigeon (Snares Cape Pigeon and Southern Cape Pigeon – above), from which Antarctic Petrel can readily be distinguished by lacking the chequerboard upperwing pattern. **Status** Uncommon visitor. **Notes** Occasionally wrecked in large numbers on mainland New Zealand beaches.

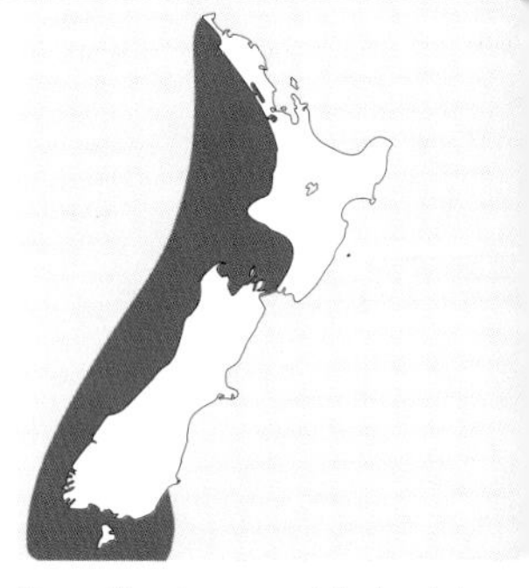

Range Breeds on nunataks (peaks), cliffs and islands in the Antarctic region. Strays to mainland New Zealand waters during winter.
Where to see In waters around, and to south of, subantarctic islands.

Snares Cape Pigeon / Titore (left) and Southern Cape Pigeon / Titore

Antarctic Petrel

Antarctic Fulmar

Fulmarus glacialoides

Identification 50 cm. Medium-sized to large, gull-like petrel with white underparts, bluish-grey upperparts, darker on upperwings. White triangle at base of primaries of upperwing conspicuous in flight. Bill flesh-coloured, feet and legs pinkish-grey. **Similar species** Could be confused with gulls, but can be distinguished by petrel-like flight and at close range by tube-nosed, flesh-coloured bill with hooked tip. **Status** Locally common visitor. **Notes** Flocks of over 5000 birds have been recorded feeding around whaling ships.

Range Breeds in Antarctic region. Regularly strays to mainland New Zealand in winter.
Where to see Waters off southern and western New Zealand, especially during winter and spring.

© Don Hadden

Antarctic Fulmar

Southern Giant Petrel / Pangurunguru

Macronectes giganteus

Identification 90 cm. Very large mollymawk-sized petrel with thickset body, greyish-brown, often with white feathers on crown, face, throat and chest. In older birds, leading edge of wings is often noticeably paler in colour. Massive horn-coloured bill with greenish tip and prominent nostril tube. All-white morphs occur. **Similar species** Northern Giant Petrel (below), from which it differs by being somewhat lighter in colour and by having a white morph. Bill tip greenish in Southern Giant Petrel and brownish in Northern Giant Petrel. **Status** Common visitor. **Notes 1.** Often congregates around fishing vessels. **2.** Formerly congregated at outfalls of freezing works and was once common around whaling stations.

Range Breeds on Antarctic coast and circumpolar subantarctic islands. Common visitor to New Zealand coastal waters.
Where to see New Zealand coastal waters, especially in winter.

Northern Giant Petrel / Pangurunguru

Macronectes halli

Identification 90 cm. Very large mollymawk-sized petrel with thickset body, greyish-brown with paler feathers on face, throat and chest. Massive horn-coloured bill with brown to reddish-brown tip and prominent nostril tube. **Similar species** Southern Giant Petrel (above), from which it differs by being somewhat darker in colour and by lacking a white morph. Bill tip brownish in Northern Giant Petrel and greenish in Southern Giant Petrel. **Status** Common native. **Notes 1.** Often congregates around fishing vessels. **2.** Formerly congregated at outfalls of freezing works and around whaling stations.

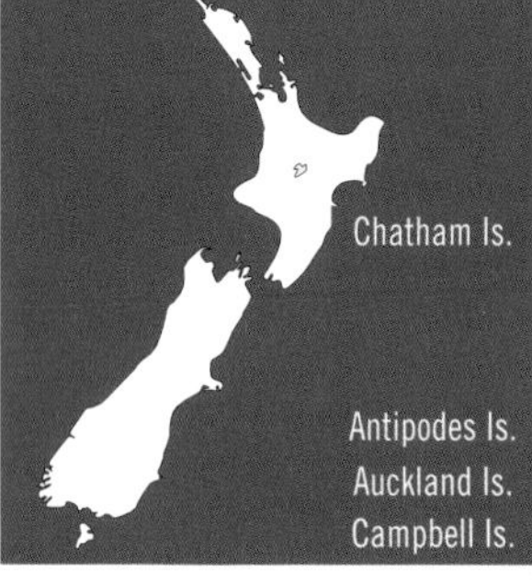

Range Breeds at the south of Stewart Island as well as on many subantarctic islands. Common winter visitor to coastal waters around New Zealand, and may enter some distance into harbours, following vessels.
Where to see New Zealand coastal waters, especially in winter.

Southern Giant Petrel / Pangurunguru

Northern Giant Petrel / Pangurunguru

Fairy Prion / Titi-wainui

Pachyptila turtur

Identification 25 cm. The smallest prion. Blue-grey upperparts, white underparts, broad black tip to tail, bold black 'M' shape across upperwings. Stout blue bill. **Similar species** Other prions, from which Fairy may be distinguished by its small size, short bill, and broad black tail tip. **Status** Locally abundant native. **Notes 1.** New Zealand's most common prion and most abundant beach-wrecked tubenose species. **2.** Does not follow vessels but is sometimes attracted by lights of stationary ships at night.

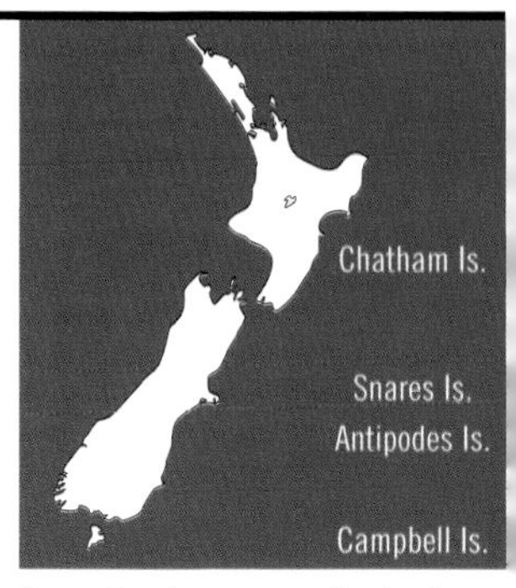

Range Breeds on many subantarctic islands. In New Zealand region, breeds on many rodent-free offshore islands from Poor Knights south, and at Snares and Chathams. Ranges widely in New Zealand coastal waters and north into subtropical seas.

Where to see Coastal and offshore waters from Northland's east coast south to Foveaux Strait and around the Chatham Islands.

Fulmar Prion

Pachyptila crassirostris

Identification 26 cm. Blue-grey upperparts, white underparts, broad black tip to tail. Bold black 'M' shape across upperwings. Stout blue bill with a large nail. **Similar species** Other prions, especially Fairy Prion (above), which is darker with a slightly smaller bill. **Status** Locally common native. **Notes** Probably the rarest prion on a global scale.

Range Breeds in the South Indian Ocean and New Zealand regions. Main New Zealand breeding sites are at Chatham, Bounty, Snares and Auckland Islands. Probably ranges widely in southern New Zealand waters.

Where to see Most likely to be seen in seas south of Stewart Island and around the Chatham Islands.

Fairy Prion / Titi-wainui

Fulmar Prion

Thin-billed Prion

Pachyptila belcheri

Identification 26 cm. Blue-grey upperparts, white underparts and with a paler face and more prominent white eyebrow than other prions. 'M' pattern on upperwings less distinct than other prions, and it has a narrow black tip to the tail. Slender bill. **Similar species** Other prions, however Thin-billed has a paler face and less distinct black markings on upperwings and tail. **Status** Common visitor.

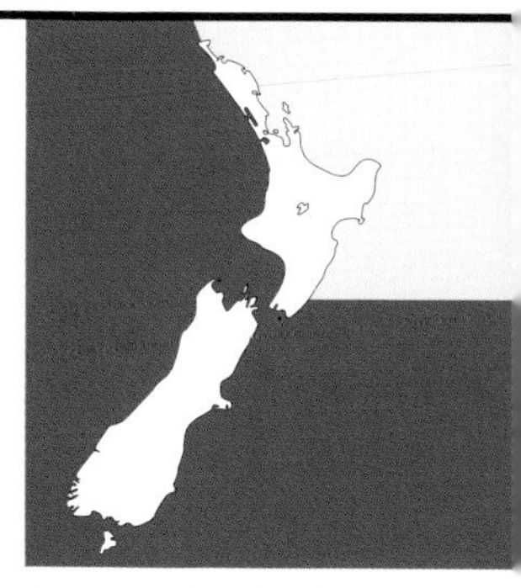

Range Breeds in the subantarctic Atlantic and Indian Oceans. Ranges widely in subantarctic waters, regularly reaching New Zealand in winter and spring, where it is usually found as a beach-wreck.

Where to see Southern and western seas around New Zealand, especially in winter and spring. Probably most easily seen as a beach-wreck.

Antarctic Prion / Whiroia

Pachyptila desolata

Identification 26 cm. Blue-grey upperparts, white underparts. Blackish 'M' pattern across upperwings and a narrower black tip to tail than Fairy and Fulmar Prions (p. 58). Stout, broad, bluish bill with no visible lamellae at base when bird in hand. **Similar species** Very similar to Salvin's Prion (p. 62), but in hand bill slightly narrower and lamellae not visible at base of closed bill. Not easily distinguished from Salvin's Prion at sea. **Status** Locally common native.

Range Breeds on a number of island groups in the Antarctic and subantarctic regions. Breeds in the New Zealand region at the Auckland Islands. A regular visitor in winter and spring to southern New Zealand waters.

Where to see Southern seas around New Zealand, especially in winter and spring.

Thin-billed Prion

Antarctic Prion / Whiroia

Salvin's Prion

Pachyptila salvini

Identification 27 cm. Blue-grey upperparts, white underparts. Blackish 'M' pattern across upperwings and a narrower black tip to tail than Fairy and Fulmar Prions (p. 58). Stout, broad, bluish bill with visible lamellae at base when bird in hand. **Similar species** Other prions, especially Broad-billed Prion (below), however latter has a darker head, narrower dark tail tip and a darker, broader bill. **Status** Common visitor.

Range Breeds on subantarctic Indian Ocean Islands. Ranges east from its breeding islands as far as New Zealand in winter and spring, where it is commonly found as a beach-wreck.
Where to see Southern and western seas around New Zealand, especially in winter and spring. Most easily seen as a beach-wreck.

Broad-billed Prion / Parara

Pachyptila vittata

Identification 28 cm. Largest prion. Blue-grey upperparts, white underparts. Black 'M' pattern across upperwings, narrow black tip to tail. Darkish head with distinct white eyebrow. Very broad grey bill with clearly visible lamellae when bird in hand. **Similar species 1.** Other prions, but the Broad-billed is possibly the easiest prion to identify at sea. Distinctive features include large size and large head emphasised by the robust bill and high forehead. Flies more slowly than the smaller prions and glides more. **2.** At sea the Blue Petrel (p. 64) could be confused with Broad-billed Prion, but look for white-tipped tail in Blue Petrel. **Status** Locally common native. **Notes** Threatens the small population of Chatham Petrels through competition for breeding burrows.

Range Breeds on South Atlantic and southern New Zealand islands and the Chatham Islands. Local Broad-billed Prions range mainly through temperate waters around New Zealand, the Tasman Sea, Bass Strait and the South Pacific.
Where to see Southern New Zealand waters (e.g. Foveaux Strait) and around the Chatham Islands.

Salvin's Prion

Broad-billed Prion / Parara

Blue Petrel

Halobaena caerulea

Identification 30 cm. Small, prion-like petrel, greyish-blue above, with distinctive white-tipped tail, dark cap and dark 'M' pattern on upperwings. Bill is bluish-black, legs and feet blue with flesh-coloured webs. **Similar species** Easily confused with prions, however Blue Petrel has a white-tipped tail and is slightly bigger. **Status** Uncommon visitor. **Notes** Sometimes wrecked on mainland beaches in large numbers.

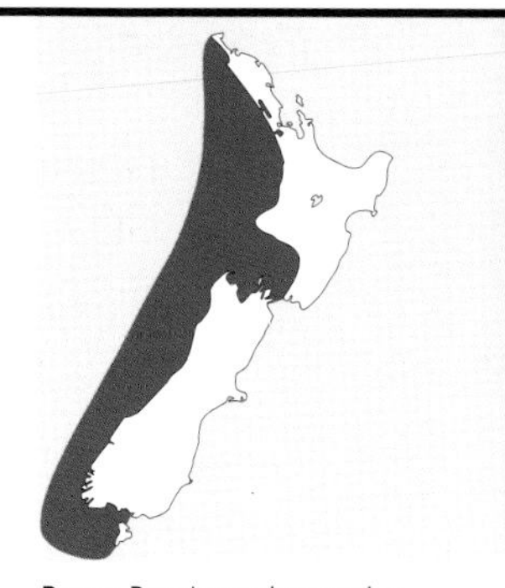

Range Breeds on circumpolar subantarctic islands. Strays to New Zealand in winter.
Where to see In subantarctic waters, but may sometimes be seen off South and Stewart Islands in winter.

Pycroft's Petrel

Pterodroma pycrofti

Identification 28 cm. Small, with pale grey upperparts, white underparts and underwings with narrow black line extending in towards body from bend of wing. Dark 'M' pattern across upperwings. Cap slightly darker than Cook's petrel with more dark feathering around eye. Bill thin and black, feet mauve with black outer toes and webs. **Similar species** Cook's Petrel (p. 66), but birds in southern New Zealand waters are most likely to be Cook's. **Status** Rare endemic. **Notes** Kiore (Pacific Rat) is a significant predator of Pycroft's Petrel on several of its breeding islands. Recent rodent eradication programmes on these islands should assist the recovery of this species.

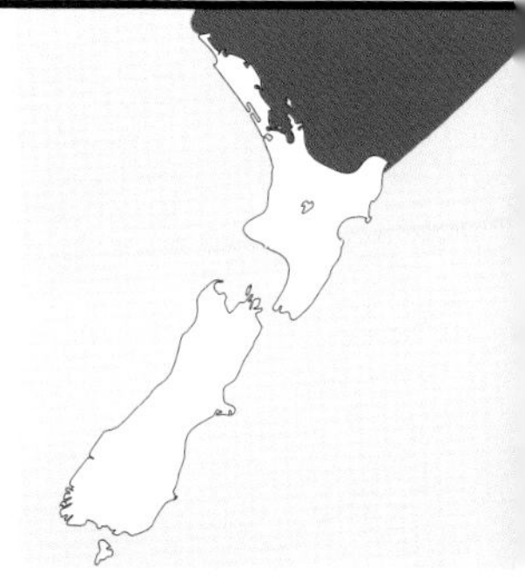

Range Breeds on Poor Knights, Mercury Islands and Hen & Chickens Group. Migrates in winter to North Pacific.
Where to see In waters off the north-eastern North Island in summer, especially around the breeding islands.

© Peter Lansley

Blue Petrel

© Don Merton

Pycroft's Petrel

Cook's Petrel / Titi

Pterodroma cookii

Identification 29 cm. Small, with pale grey upperparts, white underparts and white underwings with narrow black line extending in towards body from bend of wing. Head grey with mottled forehead. Dark 'M' pattern across upperwings. Tail dark-tipped. Bill thin and black, feet mauve with black outer toes and webs. **Similar species** **1.** Black-winged Petrel (below), from which Cook's can be distinguished by thinner black line extending in from bend of wing. **2.** Practically indistinguishable from Pycroft's Petrel (p. 64), but birds seen off southern New Zealand coasts are most likely to be Cook's Petrels. **3.** **Status** Uncommon endemic. **Notes 1.** The removal of cats from Little Barrier Island and Weka from Codfish Island has assisted recovery of the Cook's Petrel population. **2.** Formerly bred on mainland New Zealand.

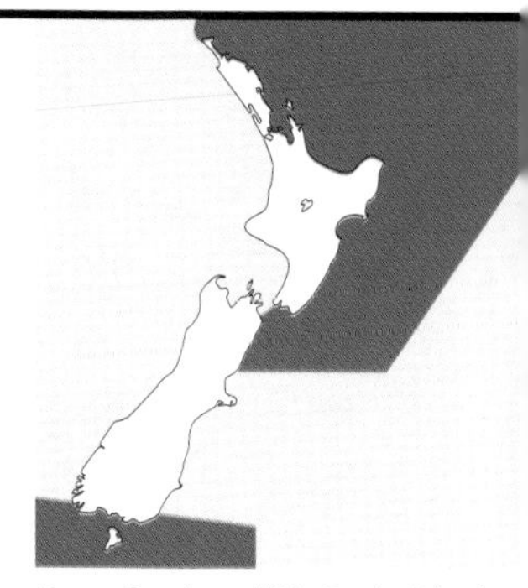

Range Breeds on Little Barrier Island in the north and Codfish Island (off Stewart Island) in the south. Migrates in winter to the eastern tropical and subtropical Pacific.
Where to see In waters of the Hauraki Gulf especially around Little Barrier Island and off Stewart and Codfish Islands in summer.

Black-winged Petrel

Pterodroma nigripennis

Identification 30 cm. Small, with dark grey upperparts and mottling on the forehead and dark patch below eye. Dark 'M' pattern across upperwings. Underparts and underwings white with diagnostic bold black marking extending in towards body from bend of wing. **Similar species** Cook's Petrel (above) and Pycroft's Petrel (p. 64), from which Black-winged can be distinguished by bold markings on underwings. **Status** Locally common native. **Notes 1.** Expanding its range, and has been found prospecting breeding sites on some mainland headlands in the far north of Northland. **2.** In the last few decades has also colonised a number of additional islands in the south-west Pacific.

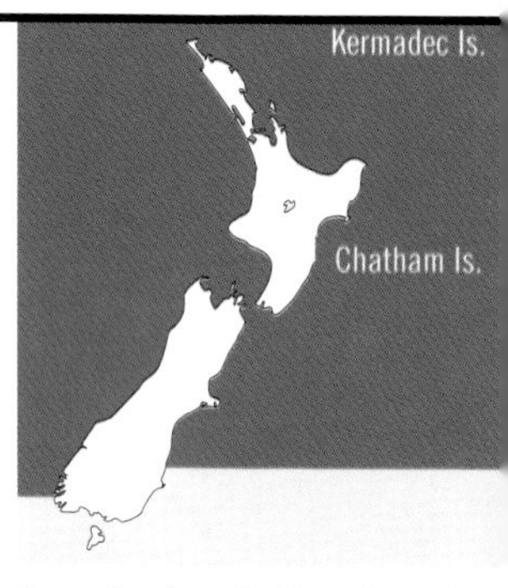

Range Breeds on the Kermadec and Three Kings Islands, East Island (off East Cape), Portland Island (off Mahia Peninsula), and on the Chatham Islands. In winter migrates to the North Pacific.
Where to see Coastal waters, usually well offshore, of the northern North Island; around Chatham Islands in summer.

Cook's Petrel / Titi

Black-winged Petrel

Chatham Petrel / Ranguru

Pterodroma axillaris

Identification 30 cm. Small, with grey upperparts and neck, and obvious dark mark below eye. Dark 'M' pattern across upperwing. Underparts white with diagnostic broad, black band extending from bend of wing to the body. **Similar species** Black-winged Petrel (p. 66), but underwing of Chatham Petrel is much bolder and extends to body. **Status** Very rare endemic. **Notes 1.** Very rare: total breeding population probably around 100 pairs. **2.** Broad-billed Prions (p. 62) compete for nesting burrows and may kill chicks and adult birds.

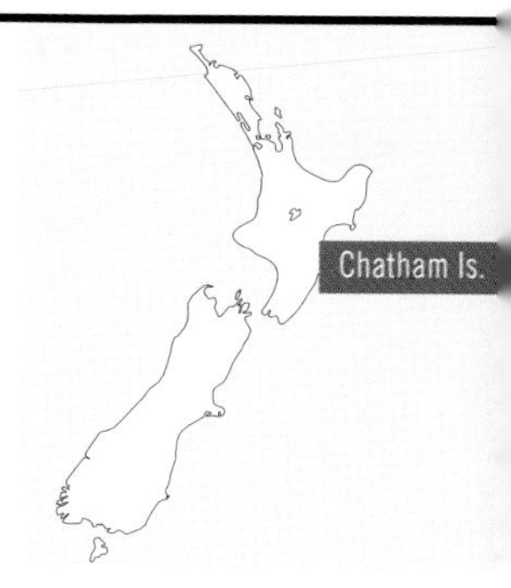

Range Breeds only on Rangatira (South-East) Island in the Chathams. Probably migrates to North Pacific in winter.
Where to see Coastal waters of the southern Chatham Islands.

Mottled Petrel / Korure

Pterodroma inexpectata

Identification 34 cm. Medium-sized gafly petrel with mottled grey upperparts and broad 'M' pattern across upperwings. Underparts white with diagnostic large, grey patch on belly and bold black, crescent-shaped mark extending inwards from bend of the wings on underwing. Bill black, legs and feet pink with black toes. **Similar species** Can be distinguished from other grey and white gadfly petrels by large dark patch on belly and bold patterns on underwings. **Status** Locally common endemic. **Notes 1.** Formerly bred on the North and South Island mainland, but wiped out by forest clearance and predators. **2.** Many Mottled Petrels on Codfish Island (off Stewart Island) once fell victim to Weka, which have now been removed.

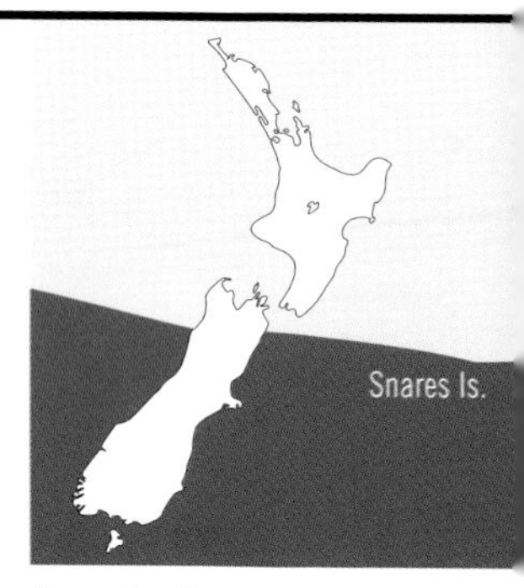

Range Breeds on islands around Fiordland and Stewart Island, and Snares Islands. Migrates to the North Pacific in winter.
Where to see In waters south of about 45° south in summer.

Chatham Petrel / Ranguru

Mottled Petrel / Korure

White-naped Petrel

Pterodroma cervicalis

Identification 43 cm. Large gadfly petrel with grey upperparts and prominent dark 'M' pattern across upperwings. Broad white nape band contrasts strongly with black cap and is diagnostic. Legs and feet pink with black toes. Like the smaller grey and white gadfly petrels, has a black line extending in from bend of wing on underwing. **Similar species** Can be distinguished from other local petrels by large size and distinctive white nape. **Status** Uncommon endemic. **Notes 1.** Closely related to Juan Fernandez Petrel, *P. externa*. **2.** Wiped out by feral cats and Norway Rats from its main Kermadec group breeding ground at Raoul Island by 1970, but discovered breeding on Macauley Island in 1969. **3.** A few now breed on Philip Island off Norfolk Island.

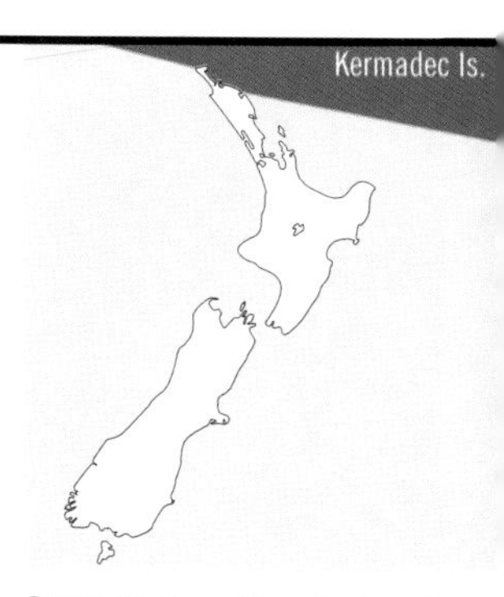

Range Breeds on Macauley Island in the Kermadec group. Occasionally strays to waters off Northland. Migrates to north-west Pacific in winter.
Where to see Waters around the Kermadec Islands and north to Fiji and Tonga.

Kermadec Petrel / Pia Koia

Pterodroma neglecta

Identification 38 cm. Medium-sized, with several colour phases. Most common is a uniform brownish-black except for diagnostic pale triangular patches at the bases of the primaries on the underwings. Corresponding pale shafts visible on primaries on upperwings. Intermediate phase is dark-headed with pale belly, while light phase is ashy grey-brown with pale face and head. Intermediate and light phases have same wing patches described for dark phase. Bill black, legs and feet vary in colour from pale flesh to black. **Similar species** Can be distinguished from most other petrels by distinctive white wing patches. **Status** Locally common native. **Notes 1.** Once bred in large numbers on Raoul Island in the Kermadecs but was eradicated there by feral cats and rats. **2.** Proposed eradication programme may allow this species to recolonise Raoul from the nearby Meyer Islands.

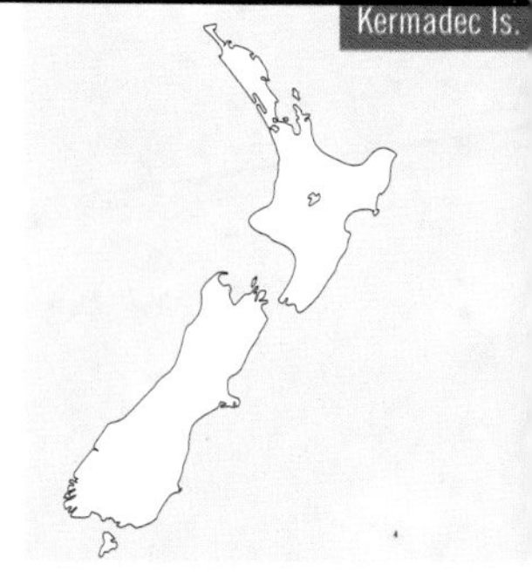

Range Breeds on the Kermadec Islands. Occasionally strays to mainland New Zealand waters. In winter migrates to the tropical Pacific. Also occurs widely on many island groups in the Tasman Sea and subtropical Pacific.
Where to see Waters around the Kermadec Islands.

© Rod Morris

White-naped Petrel

© Gary Melville

Kermadec Petrel / Pia Koia

Grey-faced Petrel / Oi

Pterodroma macroptera

Identification 41 cm. Large, almost entirely blackish-brown apart from grey face. Wings long and narrow. Short, stout, black bill. Legs and feet black. Spectacular flight, with sweeping arcs and much soaring. **Similar species 1.** Black Petrel (p. 48), which is larger and blacker, also lacks grey face and has a pale bill. **2.** Flesh-footed Shearwater (p. 38), which is browner, has a pinkish bill and flesh-coloured feet and slow-flapping flight. **Status** Common native. **Notes 1.** Breeds during winter. **2.** Once bred on many headlands and clifftops on mainland New Zealand, but now eliminated from most of these by introduced predators. **3.** The main muttonbird harvested by northern Maori, with the young taken in November.

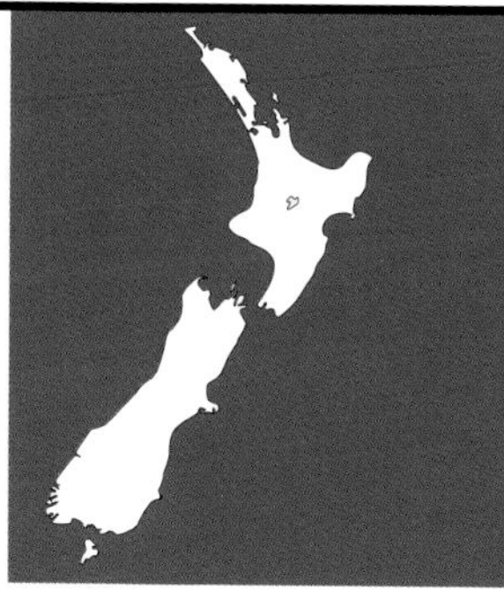

Range In New Zealand, breeds on many islands around the northern North Island, and on some mainland headlands. New Zealand birds range widely in the south-west Pacific and Tasman Sea. Also occurs off Western Australia, and the South Indian and South Atlantic Oceans, where it is known as the Great-Winged Petrel.
Where to see Coastal waters, especially off northern North Island.

Chatham Island Taiko / Magenta Petrel

Pterodroma magentae

Identification 38 cm. Medium-sized gadfly petrel. Sooty grey above with darker head. Underwings and throat also sooty grey, contrasting sharply with white underparts. Stout black bill. Legs and feet pink with dark outer toes and webs. **Similar species 1.** May be distinguished from other petrels by sooty grey head and underwings, which contrast with the white underparts. **2.** Rather similar colour pattern to some tropical species such as Phoenix and Tahiti Petrels and intermediate phase Kermadec Petrel (p. 70) but lacks the white wing patches of the last species. **Status** Very rare endemic. **Notes** Long presumed extinct, but rediscovered on the Chatham Islands in 1979. Chatham Island Taiko is now the centre of an intensive conservation management project.

Range Known to breed only on main island (Chatham) of the Chatham Islands. Ranges throughout waters to south and east of Chathams. Presumed to migrate in winter to tropical Pacific.
Where to see Best chance is in waters to south and east of Chatham Islands in summer, or by joining one of the Taiko management expeditions to the main Chatham Islands.

Grey-faced Petrel / Oi

Chatham Island Taiko / Magenta Petrel

White-headed Petrel

Pterodroma lessonii

Identification 43 cm. Large, with distinctive white head and tail. Grey back and mantle with dark 'M' pattern across upperwings. Underparts white, with underwings greyish-black. Conspicuous black mark around eye. Bill black and sturdy. Legs and feet pinkish with dark tips to toes. **Similar species** Grey-faced Petrel (p. 72), from which White-headed can be distinguished by its white head and tail and white belly. **Status** Locally common native. **Notes 1.** Skuas prey on large numbers of White-headed Petrels on some of their breeding islands. **2.** Usually ignores vessels.

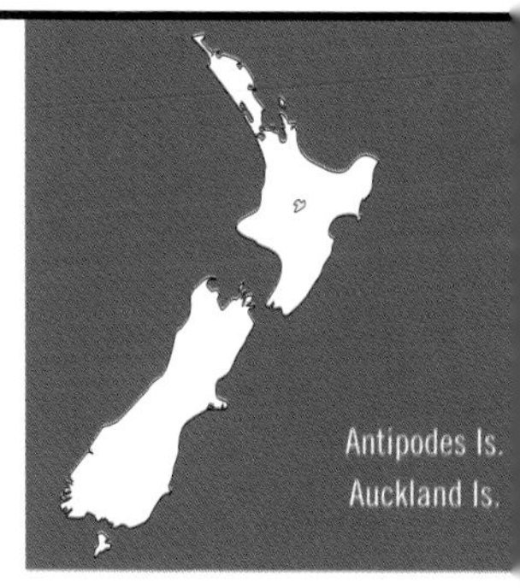

Range Breeds on Auckland and Antipodes Islands. Also breeds on Macquarie Island and south Indian Ocean islands. Occasionally seen off the coast of mainland New Zealand during winter. Ranges between the latitudes of New Zealand and the edge of the Antarctic pack ice.
Where to see In waters between Stewart Island and the pack ice.

Soft-plumaged Petrel

Pterodroma mollis

Identification 34 cm. Medium-sized gadfly petrel, ashy-grey upperparts with darker upperwings. Underwings dark, underparts white with grey band across the chest. Bill black, legs and feet pinkish with dark tips to toes. **Similar species** Can be distinguished from other petrels by size, grey underwings, and grey band across chest. **Status** Uncommon native. **Notes** Once rare but now visiting mainland New Zealand waters in increasing numbers.

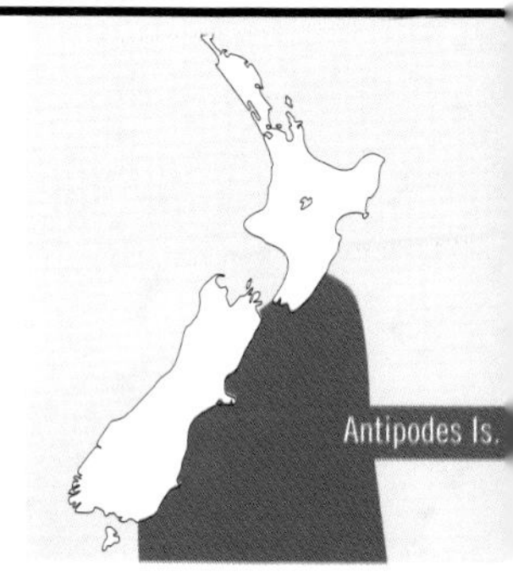

Range Breeds on Antipodes Islands. Elsewhere, breeds widely on sub-antarctic Atlantic and Indian Ocean islands.
Where to see Occasionally in waters off southern and south-eastern New Zealand.

White-headed Petrel

Soft-plumaged Petrel

Wilson's Storm Petrel

Oceanites oceanicus

Identification 18 cm. Starling-sized, sooty-brown petrel with conspicuous white rump, square tail and pale wing-bar. Legs black with noticeable yellow-webbed feet projecting beyond tail. When not flying erratically, skipping on surface of sea, flight fast and direct, interspersed with long glides. **Similar species 1.** White-bellied Storm Petrel (p. 80) and Black-bellied Storm Petrel (p. 78), from which Wilson's can be distinguished by its dark underparts. **2.** Leach's Storm Petrel (a very rare vagrant to New Zealand waters) is distinguished by its white rump patch which has greyish line up the centre. **Status** Locally common native. **Notes 1.** Occasionally follows vessels. **2.** Possibly one of the world's most abundant bird species.

Range Breeds on Antarctica and some more southern subantarctic islands. Sometimes seen on passage through New Zealand waters to the Northern Hemisphere in November–December and March–May.
Where to see Occasionally in coastal and offshore waters in November–December, and March–May.

Grey-backed Storm Petrel / Reoreo

Oceanites nereis

Identification 18 cm. Very small storm petrel with dark upperwings, pale grey rump and square, dark-tipped tail. Dark head and throat contrast sharply with white underparts. **Similar species** White-bellied Storm Petrel (p. 80) and Black-bellied Storm Petrel (p. 78), but Grey-backed is smaller with paler underparts and grey, not white, rump. **Status** Locally common native. **Notes 1.** Usually solitary at sea. **2.** Nests on surface under tussocks and other vegetation. Absent from islands where introduced predators are present.

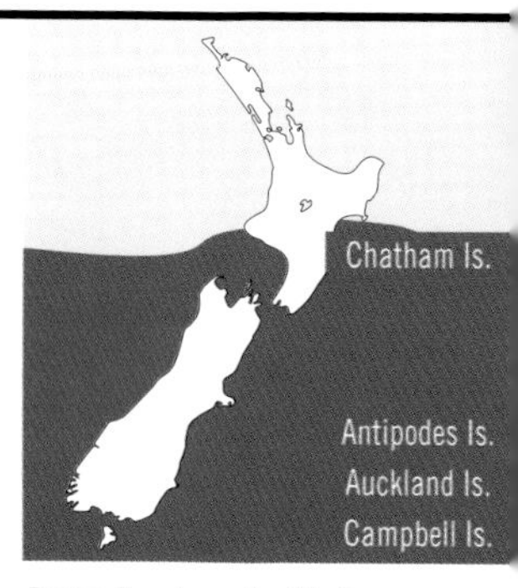

Range Breeds on the Chatham, Auckland, Antipodes and Campbell Islands. Sedentary in seas around breeding islands, but does range into subtropical seas north of New Zealand.
Where to see In seas around Chatham Islands, south of Stewart Island and in subantarctic waters.

Wilson's Storm Petrel

Grey-backed Storm Petrel / Reoreo

White-faced Storm Petrel / Takahi-kare-moana

Pelagodroma marina

Identification 20 cm. Blackbird-sized storm petrel with white face and eyebrow, grey crown and conspicuous dark eye-stripe. Greyish-brown upperparts, with paler rump, primaries almost black. Underparts white. Yellow-webbed feet project beyond the slightly forked, dark tail. **Similar species** Can be distinguished from all other storm petrels by its white face and underparts and dark eye-stripe. **Status** Common native. **Notes 1.** Does not usually follow vessels. **2.** Sometimes encountered in small flocks of up to 20 birds.

Range Breeds at a number of colonies on predator-free islands off the main islands of New Zealand, and on Kermadec, Auckland and Chatham Islands.
Where to see Coastal waters in summer; common in the Hauraki Gulf.

Black-bellied Storm Petrel / Takahi-kare

Fregetta tropica

Identification 20 cm. Robust blackbird-sized storm petrel usually with blackish-brown upperparts, distinctive white rump and variable black stripe down the centre of the belly. Bill and feet black, feet extend beyond square tail. **Similar species** White-bellied Storm Petrel (p. 80), which lacks black belly stripe and has shorter legs that do not extend beyond tail. **Status** Locally common native. **Notes 1.** Occasionally follows vessels. **2.** Migrates to tropical South Pacific in winter.

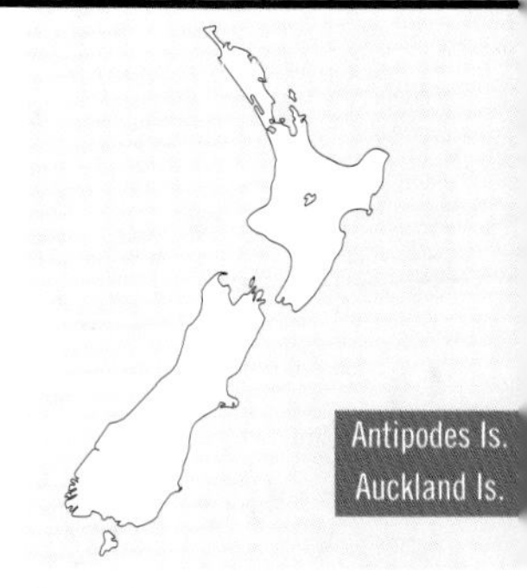

Range Circumpolar subantarctic distribution. In New Zealand region breeds on Auckland and Antipodes Islands. Sometimes seen on passage to the tropical South Pacific in about May and on return in October–November.
Where to see Sometimes encountered well offshore in oceanic waters in May and October–November.

White-faced Storm Petrel / Takahi-kare-moana

Black-bellied Storm Petrel / Takahi-kare

White-bellied Storm Petrel

Fregetta grallaria

Identification 20 cm. Robust blackbird-sized storm petrel, usually with blackish-brown head and upperparts, distinctive white rump and white belly. Bill and feet black, feet do not extend beyond square tail. **Similar species** Black-bellied Storm Petrel (p. 78), which usually has black belly stripe, and has longer legs. **Status** Rare native. **Notes** Little known of its breeding habits.

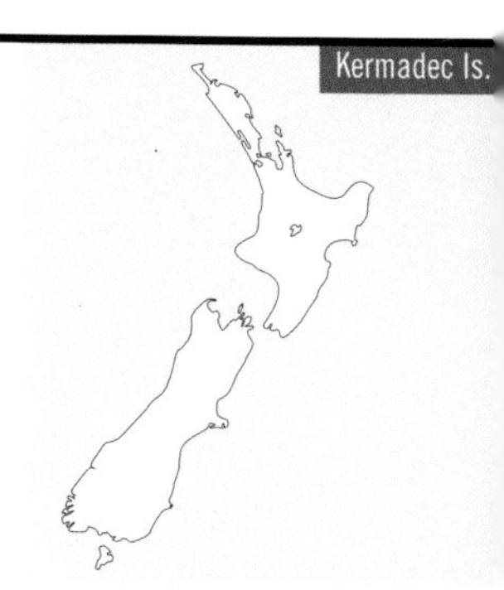

Range Breeds on many islands in subtropical Atlantic, Indian and Pacific Oceans. Near New Zealand, breeds on Lord Howe Island and Kermadec Islands.
Where to see In the north Tasman and the seas north of New Zealand and around the Kermadec Islands.

White-bellied Storm Petrel

Yellow-eyed Penguin / Hoiho

Megadyptes antipodes

Identification 65 cm. Medium-sized with white underparts, upperparts slaty-grey. Head pale golden-yellow with clearly defined yellow band running from eye around back of head; this feature less clearly defined in juveniles. **Similar species** At sea could be confused at a distance with Blue Penguin (p. 84) but Blue is much smaller, bluer and lacks yellow head band. **Status** Uncommon endemic. **Notes** One of New Zealand's rarest penguins, threatened by predation and human induced changes to its coastal breeding habitat.

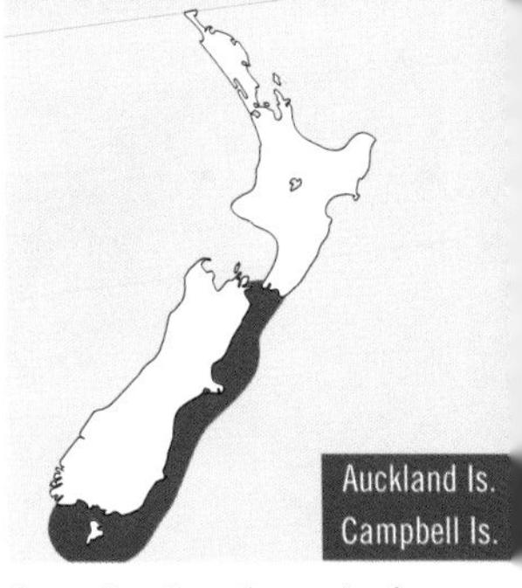

Range Breeds on the coasts of Canterbury (from Banks Peninsula southwards), Otago and Southland. Also breeds on Stewart, Auckland and Campbell Islands. In winter, may range as far north as Cook Strait.
Where to see Moeraki, Oamaru, Otago Peninsula, e.g. Taiaroa Head, Nugget Point, and around Stewart Island.

Gentoo Penguin

Pygoscelis papua

Identification 75 cm. Moderately large penguin. Slaty-grey head and throat. Upperparts flecked with white. Distinctive white triangle above eye which meets on crown. Bill, legs and feet orange. **Similar species** None. **Status** Rare subantarctic vagrant.

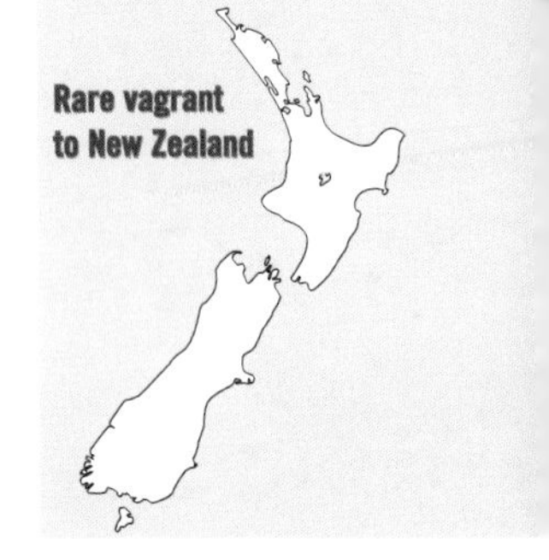

Range Breeds, in large colonies, in many areas in the circumpolar Antarctic and on Macquarie Island. Recorded as a vagrant on Campbell Island and occasionally on the Otago coast.
Where to see Subantarctic seas around Macquarie Island. A rare visitor to southern New Zealand.

Yellow-eyed Penguin / Hoiho

Gentoo Penguin

Blue Penguin / Korora

Eudyptula minor

Identification 40 cm. Small. Slaty-blue above, silvery-white below. Anterior part of flipper blue-black with white trailing edge, this feature more pronounced in the white-flippered morph found on Banks Peninsula. **Similar species** At sea could be confused at a distance with Yellow-eyed Penguin (p. 82), but Blue is much smaller, bluer and lacks yellow head band. **Status** Common native. **Notes 1.** Numbers have declined in recent years due to predation, especially by dogs, and human encroachment. **2.** The white-flippered morph from Banks Peninsula is considered by some to be a distinct species: the White-flippered Penguin, *Eudyptula albosignata*. **3.** Sometimes nests under seaside cottages, where they are noisy at night.

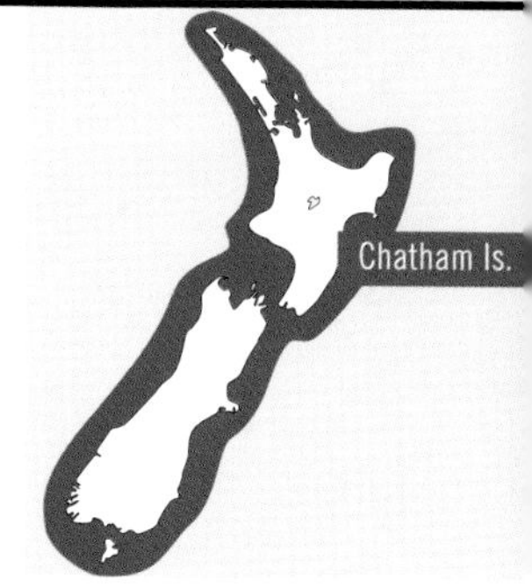

Range Breeds around coasts of mainland New Zealand and on nearer offshore islands, as well as Chatham Islands. Also found in southern Australia and Tasmania. **Where to see** Coastal waters around New Zealand. Viewing burrows at Tiritiri Matangi Island and viewing area at Oamaru Blue Penguin Colony are good places to see them. At breeding sites, birds may be seen coming ashore after dark.

Rockhopper Penguin

Eudyptes chrysocome

Identification 55 cm. Smallest of the crested penguins. Upperparts slaty blue-grey, underparts white. Has thin, golden-yellow eyebrow splaying out at back of head to form fluffy crown, with some yellow feathers drooping towards the neck. **Similar species** Other crested penguins, from which this species is most easily distinguished by smaller size and shape of crest. **Status** Locally common native. **Notes** May breed in association with Erect-crested Penguin (p. 88).

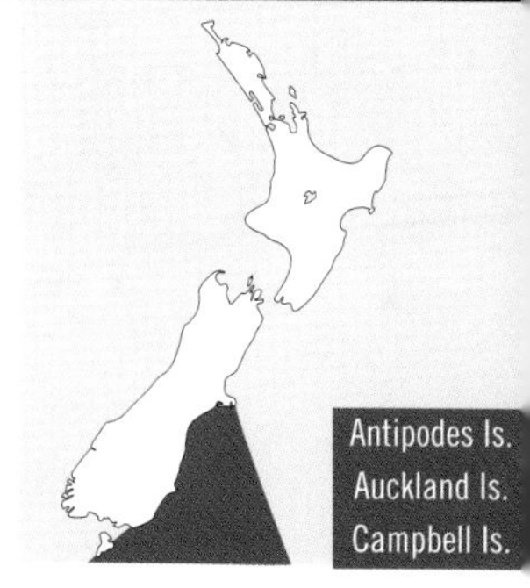

Range Breeds on many islands in the circumpolar subantarctic region. In New Zealand waters, it is mainly found on Campbell, Auckland and Antipodes Islands.
Where to see Coastal waters in the subantarctic region around breeding islands.

Blue Penguin / Korora; White-flippered morph of Blue Penguin (inset)

Rockhopper Penguin

Fiordland Crested Penguin / Tawaki

Eudyptes pachyrhynchus

Identification 60 cm. Broad yellow crest, which droops at the neck, along with five or six whitish stripes on cheek. Back slaty bluish-black, underparts white. **Similar species** Other crested penguins, from which this species is most easily distinguished by shape of its crest. **Status** Rare endemic. **Notes 1.** Has declined considerably in numbers in recent years due to predation. **2.** Probably now New Zealand's rarest penguin.

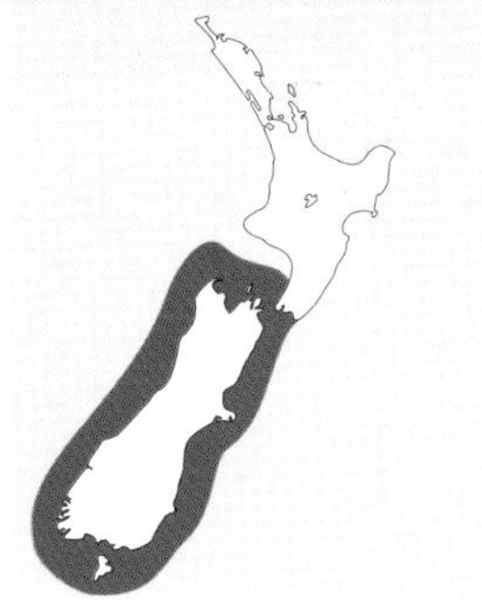

Range Breeds sparingly around the coasts of Southland, Fiordland and Westland as far north as about Haast. Also on Stewart Island and the Solanders. Seen regularly as far north as Cook Strait and occasionally ranges to south-east Australia, especially Tasmania.
Where to see West Coast of South Island, e.g. Monro Beach, Knight Point and Jackson Head.

Snares Crested Penguin / Tawaki

Eudyptes robustus

Identification 60 cm. Upperparts bluish-grey, underparts white. Thin, bright yellow eyebrow stripe which broadens and becomes bushy, forming a drooping, yellow crest behind eye. **Similar species** Other crested penguins, from which this species is most easily distinguished by shape of crest, and line of naked pink skin around base of bill. **Status** Locally common endemic.

Range Breeds only on Snares Islands. Occasionally strays to South and Stewart Islands.
Where to see Coastal waters of Snares Islands.

© DOC/A. Munn

Fiordland Crested Penguin / Tawaki

© DOC/Rod Morris

Snares Crested Penguin / Tawaki

Erect-crested Penguin

Eudyptes sclateri

Identification 60 cm. Upperparts of body slaty blue-grey, underparts white. Broad, bright yellow eyebrow stripe, which flares into a bushy, erect crest behind eye. **Similar species** Other crested penguins, from which this species is most easily distinguished by the shape of its crest. Erect-crested has broad whitish line of skin around base of bill. **Status** Locally common endemic. **Notes** Formerly bred on Campbell Island.

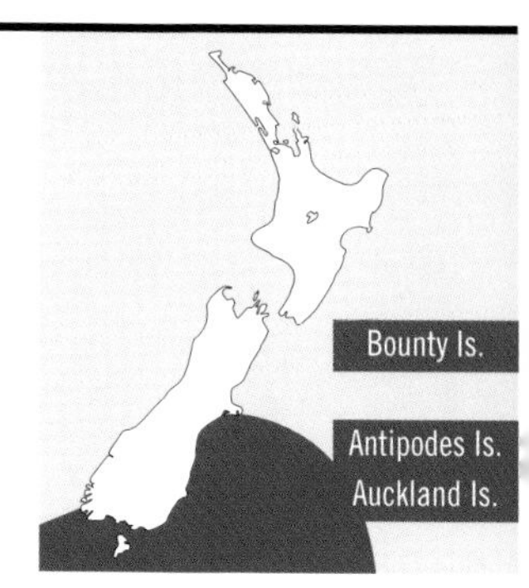

Range Breeds mainly on Bounty Islands (where it is the most abundant penguin) and Antipodes Islands and in small numbers on the Auckland Islands. Regularly strays to New Zealand waters in winter, and occasionally reaches Tasmania and southern Australia.
Where to see Coastal waters of Bounty and Antipodes Islands.

Erect-crested Penguin

Red-tailed Tropicbird / Amokura

Phaethon rubricauda

Identification 46 cm (plus up to 40 cm for tail streamers). White body with black shafts to primaries and black barring on trailing inner upperwings. A black marking through eye. Elongated, fine, red central tail feathers. Bill bright red. Juveniles white with finely barred upperparts, and dark bills. **Similar species** White-tailed Tropicbird (below), but Red-tailed is larger, has fine red tail feathers, less conspicuous black wing markings, and red bill. **Status** Rare tropical native. **Notes 1.** Does not follow vessels as regularly as White-tailed Tropicbird, but may be attracted to them. **2.** Red tail feathers were highly valued by Maori for decorative purposes.

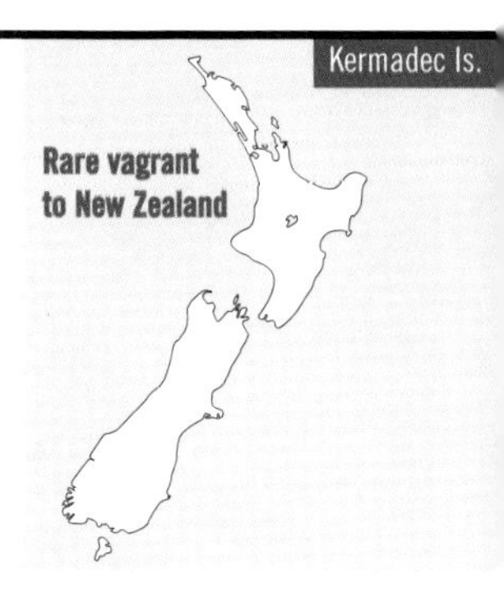

Range Breeds on many islands in tropical and subtropical Indian and Pacific Oceans. In New Zealand region, breeds on the Kermadec Islands. Occasionally seen off Northland and in the Hauraki Gulf after cyclonic weather. **Where to see** Waters off the North Island. Occasionally off Northland and in the Hauraki Gulf.

White-tailed Tropicbird

Phaethon lepturus

Identification 38 cm (plus c 40 cm for tail streamers). White body with conspicuous black markings on upperwings. A black marking through eye. Elongated white, central tail feathers. Bill yellow. Juveniles white with finely barred upperparts, less black on upperwings and pale yellow bill. **Similar species** Red-tailed Tropicbird (above), but White-tailed is smaller, has white tail feathers and conspicuous black upperwing markings. **Status** Rare vagrant. **Notes** May follow vessels.

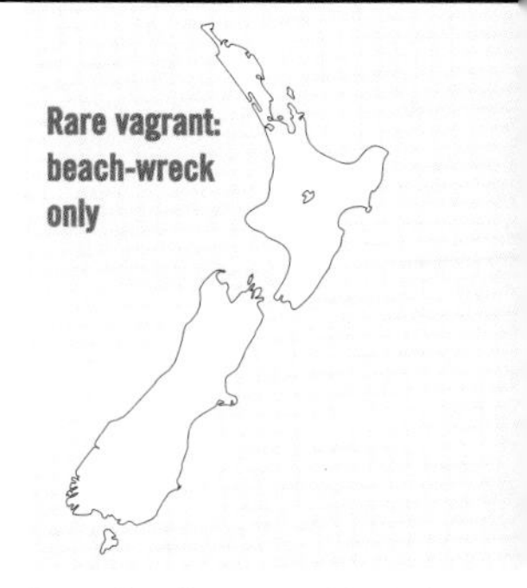

Range Breeds on many islands in tropical west Pacific, Indian and Atlantic Oceans. A rare beach-wreck around the North Island coast. **Where to see** Rarely, in seas well to the north of New Zealand.

Red-tailed Tropicbird / Amokura

White-tailed Tropicbird

Australasian Gannet / Takapu

Morus serrator

Identification 90 cm. Spindle-shaped white body with yellow head. Black flight feathers and central tail feathers. Bill bluish-grey. Long, narrow wings. **Similar species** Unlikely to be confused with any other species except for possibly the Masked Booby (p. 94), but this lacks Australasian Gannet's golden-yellow head, has yellow bill, black face and more black on wings and tail. **Status** Common native. **Notes** Flocks above schooling fish and dives spectacularly; easily seen from boats in coastal waters.

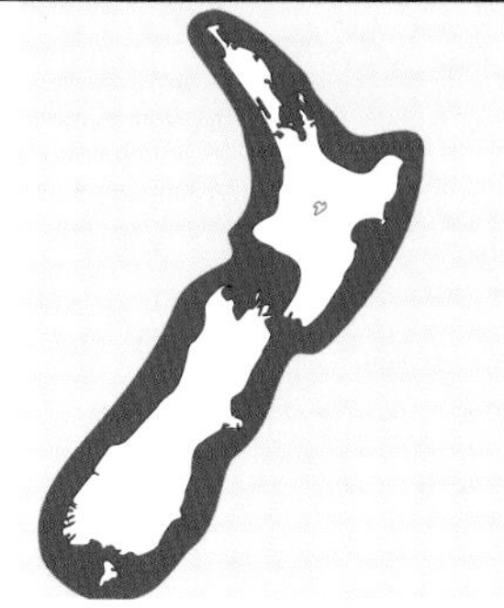

Range Breeds on many islands and some headlands around New Zealand coast. Common in coastal waters, often enters harbours and estuaries. Juveniles disperse to waters off south-east Australia, returning permanently to New Zealand when adults.
Where to see Accessible mainland colonies at Muriwai, Cape Kidnappers and Farewell Spit.

Brown Booby

Sula leucogaster

Identification 80 cm. Gannet-shaped with distinctive chocolate-brown head, neck and upperparts which contrast sharply with white underparts. Bill and feet yellowish-green. Juvenile has similar colour pattern with mottled greyish-brown belly, and greyish bill and feet. **Similar species** Could be confused with juveniles of the Masked Booby (p. 94), but juvenile Brown Booby is much browner and colour pattern of adult is evident. **Status** Rare vagrant. **Notes** A common booby in the South Pacific.

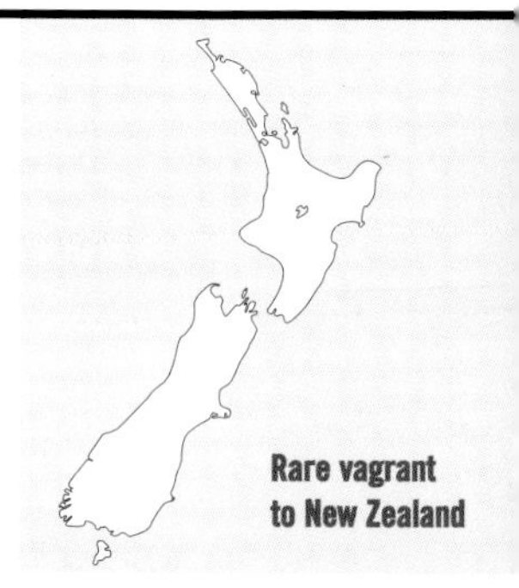

Range Breeds on many tropical islands in the Pacific, Indian and Atlantic Oceans. Regularly strays to waters of the Hauraki Gulf and Bay of Plenty.
Where to see Waters to the north of the North Island. In winter occasionally seen in Hauraki Gulf and Bay of Plenty waters.

© Dean Nixon

Australasian Gannet / Takapu

© Brian Chudleigh

Brown Booby

Masked Booby

Sula dactylatra

Identification 80 cm. Gannet-like, with white body and black trailing edge to upperwings including inner part, and black tail. Black face, yellow bill and greyish-green legs. Juveniles have brown heads and necks, and brown mottled upperparts. **Similar species** Could possibly be confused with Australasian Gannet (p. 92), but Gannet is bigger, has yellow head, lacks black face, has no black on inner trailing wing edges and only central tail feathers are black. **Status** Uncommon native. **Notes** May associate with flocks of Australasian Gannets.

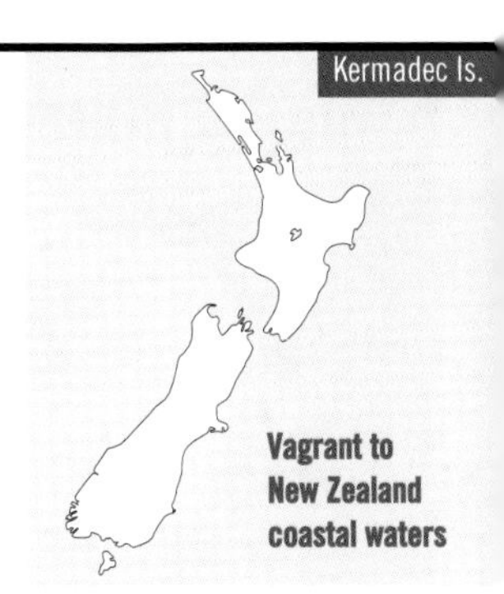

Range Breeds on the Kermadec Islands and on many other islands in tropical Indian, Pacific and Atlantic Oceans. Occasionally strays to Hauraki Gulf and Bay of Plenty waters.
Where to see Around the Kermadec Islands and seas to the north of the North Island.

Masked Booby

Black Shag / Kawau

Phalacrocorax carbo

Identification 88 cm. Large, with black body and brown wings. White patch on cheeks and throat and yellow facial skin. Breeding adults have distinctive white thigh patch. Bill grey, feet black. **Similar species** Little Black Shag (p. 98), but Little Black smaller, has dark facial skin and faster wingbeats. **Status** Common native. **Notes 1.** New Zealand subspecies is *P. c. novaehollandiae*. **2.** Wary, does not usually allow a close approach.

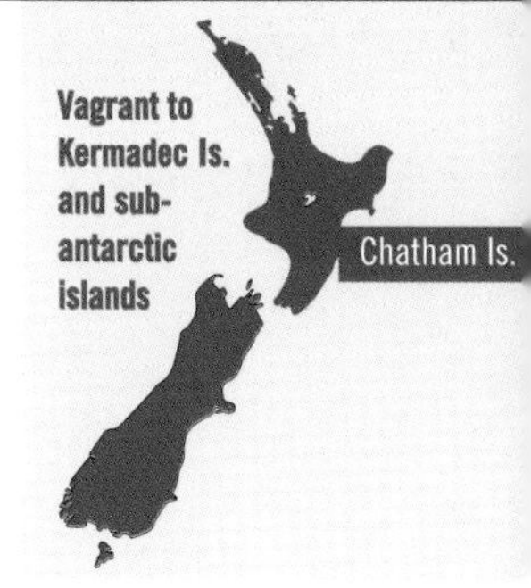

Range A cosmopolitan species ranging from eastern North America through Eurasia, Africa and Australasia. Inhabits sheltered coastal waters and inland lakes and rivers. **Where to see** Throughout New Zealand in sheltered coastal waters, lakes and rivers.

Pied Shag / Karuhiruhi

Phalacrocorax varius

Identification 81 cm. Large, with glossy black upperparts and underparts white except for black thighs. Facial skin yellowish-orange, bill flesh-coloured, feet black. **Similar species** Pied phase of Little Shag (p. 98), from which Pied can be distinguished by larger size, lack of crest, more slender flesh-coloured bill, and black thighs. **Status** Locally common native. **Notes 1.** New Zealand subspecies is *P. v. varius*. **2.** Nests and roosts in colonies, often in dead trees. **3.** Does not range very far inland from coast.

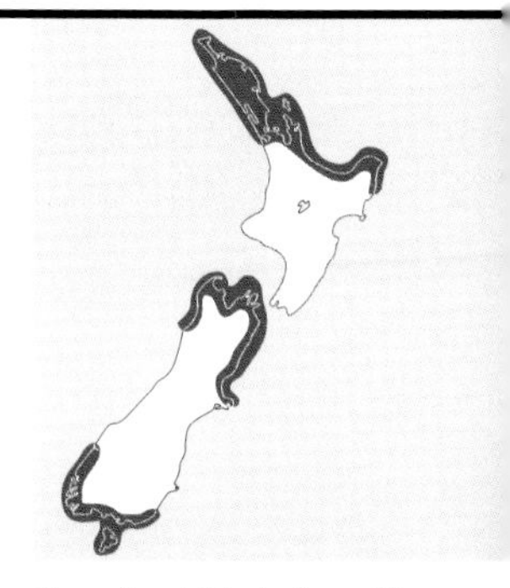

Range Coastal Australia and New Zealand. In New Zealand, most numerous in coastal Northland, Auckland and Bay of Plenty, coastal Nelson, Marlborough and North Canterbury, and Fiordland, Southland and Stewart Island. **Where to see** Coastal Northland, Auckland and Bay of Plenty, coastal Nelson, Marlborough and North Canterbury and Fiordland, Southland and Stewart Island.

Black Shag / Kawau

Pied Shag / Karuhiruhi

Little Black Shag / Kawau-tui

Phalacrocorax sulcirostris

Identification 61 cm. Small, slender, glossy black with blackish bill and black feet. Dark facial skin with conspicuous green eye. Flies rapidly, often in long lines or in V-formation. Often seen in large flocks. **Similar species** Black Shag (p. 96), but the Little Black is smaller, lacks white patches, has black facial skin, faster wingbeats, and is often seen in flocks. **Status** Locally common native. **Notes 1.** Nests in large colonies, often with Little Shags (below). **2.** Leapfrogging feeding flocks are a feature of some northern harbours in winter. **3.** Common around Auckland.

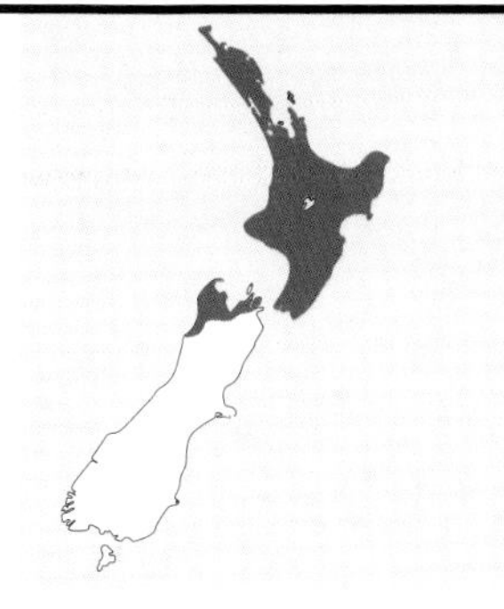

Range Lakes, estuaries and harbours of the North Island, rare in the South Island, where it mainly occurs in the Nelson and Marlborough districts.
Where to see Lakes and sheltered coastal waters, mainly in the North Island.

Little Shag / Kawau-paka

Phalacrocorax melanoleucos

Identification 56 cm. Small, with variable plumages, ranging from almost entirely dark, as in juveniles, through white-throated to fully pied. Short, yellow bill. Feet black. Adults have small crest on forehead. **Similar species 1.** Pied Shag (p. 96), but Pied is much bigger, has a more slender flesh-coloured bill, black flank marks, and lacks variable plumages. **2.** Little Black (above) similar to all-dark juvenile Little Shag, but Little Black has long, thin, greyish bill, no crest and shorter tail. **Status** Common native. **Notes 1.** New Zealand subpecies is *P. m. brevirostris*. **2.** Nests and roosts in colonies, often with other black-footed shag species.

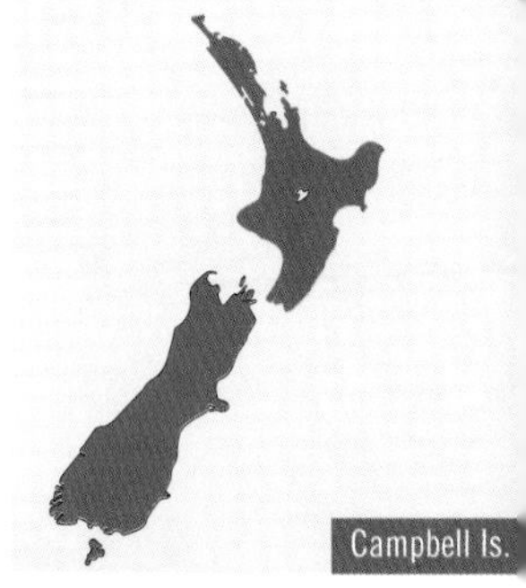

Range Throughout mainland New Zealand, Stewart Island and nearer offshore islands. Sheltered coastal waters and lakes, streams, rivers and farm ponds. Also occurs in Australia, South-east Asia and the western South Pacific.
Where to see Lakes, ponds, rivers, estuaries and sheltered coastal waters

Little Black Shag / Kawau-tui

Little Shag / Kawau-paka

King Shag / Kawau-pateketeke

Phalacrocorax carunculatus

Identification 76 cm. Large black and white shag with conspicuous white patch on upperwing and pink feet. Conspicuous blue eye ring and orange-yellow caruncles above base of bill, visible at close range. **Similar species** Pied Shag (p. 96), but King Shag smaller with white patches on upperwings and pink feet. **Status** Rare. **Notes 1.** Total population only around 500 birds. **2.** Very wary at breeding colonies, which are easily disturbed. **3.** Nests on the ground.

Range Breeds on a few small islands in outer Marlborough Sounds. **Where to see** Waters of outer Marlborough Sounds.

Stewart Island Shag / Bronze Shag / Kawau-mapua

Phalacrocorax chalconotus

Identification 68 cm. Large shag, occurs in two colour morphs which interbreed: one black and white with white wingbar; the other entirely bronzy-green. Feet pink. Conspicuous blue eye ring and orange caruncles. **Similar species** Pied Shag (p. 96), but Stewart Island Shag is smaller with white patches on upperwings and pink feet. **Status** Locally common endemic.

Range Breeds in colonies on some mainland cliffs and small islands, and ranges through coastal waters of Stewart Island, Southland and Otago. **Where to see** Coastal waters of Stewart Island, Southland and Otago.

King Shag / Kawau-pateketeke

Stewart Island Shag / Bronze Shag / Kawau-mapua

Chatham Island Shag

Phalacrocorax onslowi

Identification 63 cm. Large, like King Shag (p. 100), but has large orange caruncles. Breeding adults develop crest on forehead. **Similar species** The only large pied shag with pink feet on the Chatham Islands. **Status** Locally common endemic. **Notes** Great care needed when visiting colonies as these shags are very easily disturbed and prone to abandoning colonies as a result.

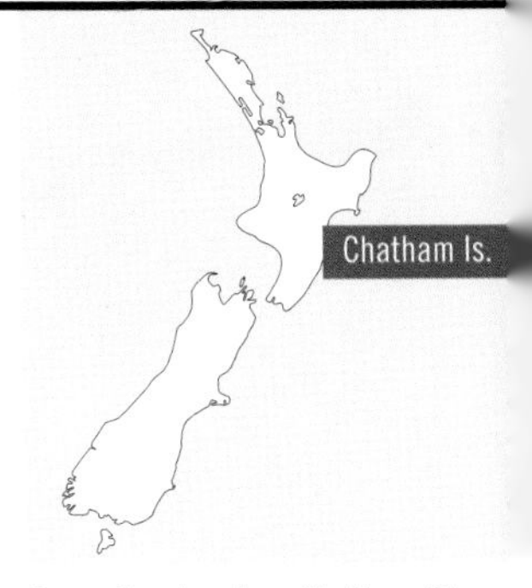

Range Breeds only on Chatham, Star Keys and Rabbit Islands in Chatham group. Ranges throughout coastal waters of Chatham Islands.
Where to see At a few breeding sites on main Chatham Island, e.g. Manukau Point, Cape Fournier, Okawa, Matarakau, and adjacent seas.

Bounty Island Shag

Phalacrocorax ranfurlyi

Identification 71 cm. Large, like King Shag (p. 100), but lacks caruncles. Breeding adults develop crest on forehead. **Similar species** The only shag on the Bounty Islands. **Status** Locally common endemic. **Notes** Breeds in colonies and feeds in flocks.

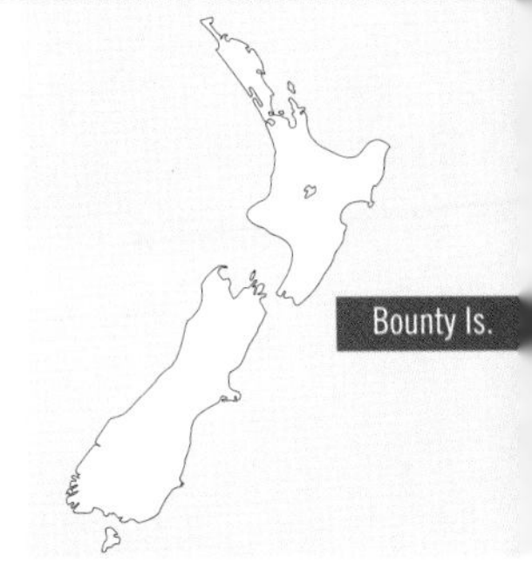

Range Confined to waters around the Bounty Islands.
Where to see Seas around the Bounty Islands.

© DOC/Rod Morris

Chatham Island Shag

© DOC/A. Booth

Bounty Island Shag

Auckland Island Shag / Koau

Phalacrocorax colensoi

Identification 63 cm. Large, like King Shag (p. 100), but lacks caruncles. Breeding adults develop crest on forehead. **Similar species** The only shag on the Auckland Islands. **Status** Locally common endemic. **Notes 1.** Nests on cliff ledges, sometimes under overhanging trees. **2.** Breeds in colonies, and feeds and roosts in flocks.

Range Confined to waters around the Auckland Islands.
Where to see Seas around the Auckland Islands.

Campbell Island Shag

Phalacrocorax campbelli

Identification 63 cm. Large, like King Shag (p. 100) but lacks caruncles and has black head and neck and white throat. Red facial skin, pink feet. Breeding adults develop crest on forehead. **Similar species** The only shag on Campbell Island. **Status** Locally common native. **Notes** Feeds in large flocks.

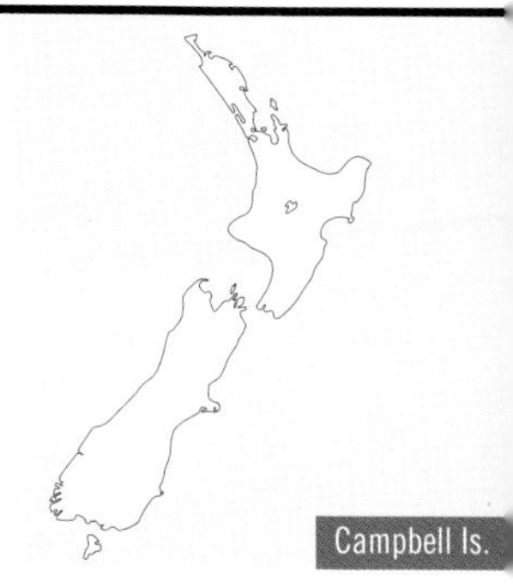

Range Sedentary, confined to waters of Campbell Island.
Where to see Waters around Campbell Island.

Auckland Island Shag / Koau

Campbell Island Shag

Spotted Shag / Parekareka

Phalacrocorax punctatus

Identification 70 cm. Slender, greyish shag with yellow feet and long, thin bill. In breeding plumage, conspicuous double crest, black face and throat, white stripe on sides of neck and black spots on upperparts and silvery underparts. Green facial skin. Non-breeding adults and immatures lack crest, and have duller plumage. **Similar species** Only mainland shag with yellow feet, greyish plumage and pale stripe on side of neck. **Status** Locally common endemic. **Notes 1.** Unmistakable nuptial plumage. **2.** In the past some colonies suffered heavy losses as a result of being shot at from boats. **3.** A cliff-nesting species. Spectacular breeding colonies in niches in cliffs around Banks Peninsula. **4.** May be seen in small numbers at Port Onehunga, Auckland during winter, with birds in spectacular breeding plumage.

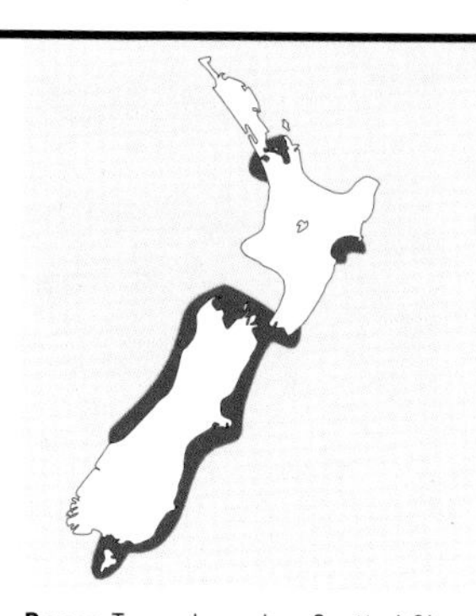

Range Two subspecies: Spotted Shag, *P. p. punctatus*, occurs locally in the North Island and northern and eastern South Island; Blue Shag, *P. p. steadi*, occurs around the south-western Stewart Island and Fiordland.
Where to see Hauraki Gulf, Auckland west coast, Wellington Harbour, most of South Island coastline and Stewart Island.

Pitt Island Shag

Phalacrocorax featherstoni

Identification 63 cm. Slender, greyish shag with dark head and neck, double crests, long thin bill and yellow feet. Black spots on upper parts, silvery-grey underparts. Green facial skin. Non-breeding adults and immatures lack crests and have duller plumage. **Similar species** Only other shags at the Chatham Islands are Chatham Island Shag (p. 102), which is black and white, and Black Shag (p. 96), which is mostly entirely black and much larger. **Status** Locally common endemic. **Notes 1.** A cliff-nesting species which nests in small colonies. **2.** Total population is c. 500 pairs.

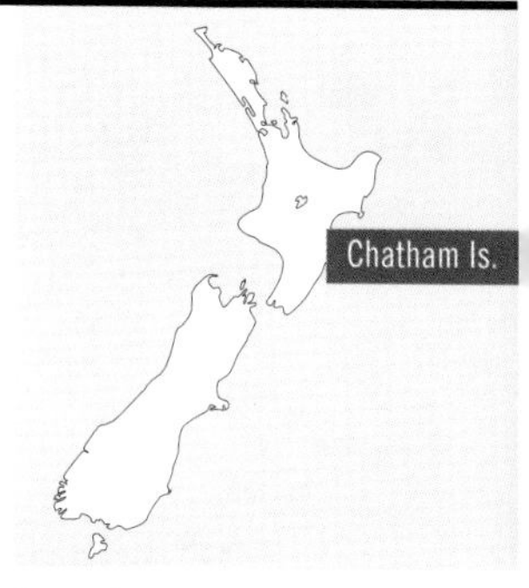

Range Coastal waters of Chatham Islands.
Where to see Headlands and cliffs around southern coast of Chatham Island, also Pitt Island and its outliers.

Spotted Shag / Parekareka

Pitt Island Shag

Greater Frigatebird / Hakuwai

Fregata minor

Identification 95 cm. Male entirely black with slight greenish sheen to upperparts and distinctive red throat patch. Female has grey throat, white chest and flanks. Juveniles have orange-brown head with white chest and flanks. **Similar species 1.** Lesser Frigatebird (below), but Greater Frigatebird males are entirely black, and in females throat greyish-white and no white on inner underwings. **2.** Juveniles similar but Greater juveniles lack white on underwings. **Status** Rare tropical vagrant. **Notes** Harries other seabirds and steals their catch and nesting materials. In rough conditions, does more of its own fishing.

Range Breeds on many islands in tropical Pacific, Indian and Atlantic Oceans. Straggles to New Zealand waters.
Where to see Seas to north of New Zealand.

Lesser Frigatebird

Fregata ariel

Identification 76 cm. Male black with diagnostic white patch stretching across chest to armpits. Female is larger, with dark head and throat, and conspicuous white patch on chest which extends to inner underwings. **Similar species** Greater Frigatebird (above), but the Lesser is smaller. Male Lesser has white patches extending onto inner underwing, while female has black throat and white chest patch that extends onto inner underwing. Juvenile has white extending onto inner underwing. **Status** Rare tropical vagrant. **Notes 1.** Like Greater Frigatebird, harries other seabirds and steals their catch. **2.** Sometimes wrecked on northern beaches after cyclones, more commonly in Northland.

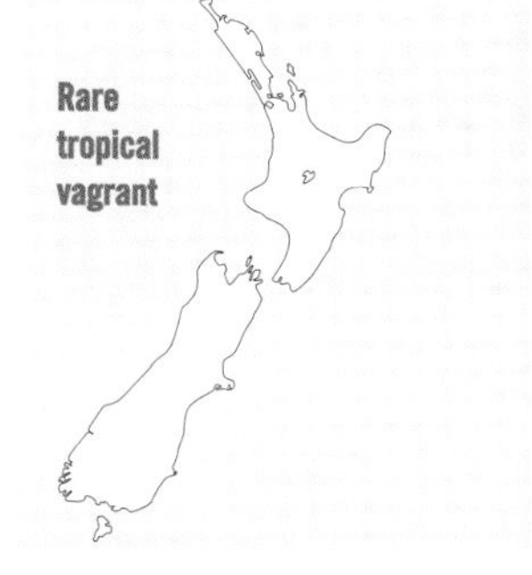

Range Breeds on many islands in tropical Pacific, Indian and Atlantic Oceans. Straggles to New Zealand waters.
Where to see Seas to north of New Zealand.

Greater Frigatebird / Hakuwai

Lesser Frigatebird

Brown Skua / Hakoakoa

Catharacta skua

Identification 63 cm. Large and sturdy, uniform dark brown apart from distinctive white patch at base of primaries. Legs and feet black as is strongly hooked bill. Powerful flight with glides. Defends territory aggressively by diving at intruders. **Similar species** Juvenile Black-backed Gull (p. 114), but Brown Skua is much larger, has darker plumage and white patch on primaries. **Status**: Locally common native. **Notes 1.** A voracious predator of other seabirds, especially smaller burrowing petrels such as prions and storm petrels. **2.** May scavenge dead farm livestock. **3.** Some nests have more than one pair in attendance.

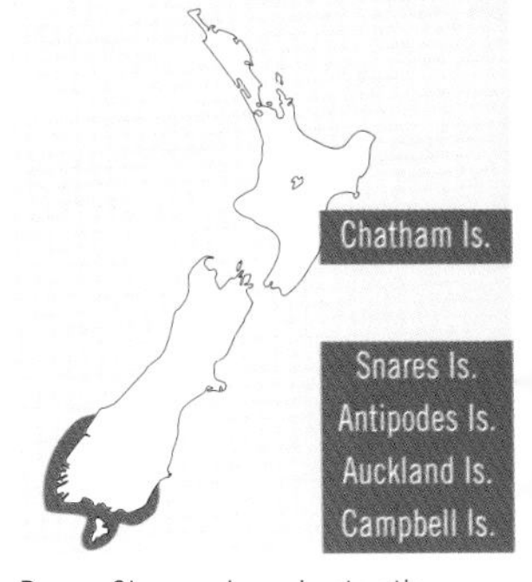

Range Circumpolar subantarctic distribution. In New Zealand region breeds on most subantarctic groups as well as on islands off coast of Fiordland, Stewart Island and Chatham Islands. Disperses throughout southern seas in winter, occasionally reaching New Zealand mainland.
Where to see Off coasts of southern South Island, Stewart Island and subantarctic islands.

South Polar Skua

Catharacta maccormicki

Identification 59 cm. Variable plumages ranging from pale honey-coloured to all dark. There are noticeable gold hackles on the nape. All have conspicuous white wing patches. **Similar species** Brown Skua (above), but South Polar is smaller, has golden nape and usually paler body colour. **Status** Occasional annual visitor. **Notes** A ruthless predator of penguin chicks and eggs at their breeding sites.

Range Breeds in Antarctica. Passes through New Zealand coastal waters in late summer on migration to North Pacific.
Where to see Occasionally, in coastal waters in late summer.

Brown Skua / Hakoakoa

South Polar Skua

Arctic Skua

Stercorarius parasiticus

Identification 43 cm. Small, angular skua with two main plumage phases. Most common dark phase blackish-brown above with yellowish tinge to nape. Conspicuous white patch on base of primaries. Pale phase has greyish-white underparts, prominent broad brown band on chest and similar wing pattern to dark phase. There are also many intermediate plumages between dark and pale phases. Juveniles have variable plumages, but usually with some mottling and barring. Breeding adults have elongated, pointed central tail feathers. **Similar species** Pomarine Skua (below), but Pomarine has two pale patches at base of primaries and breeding adults have two elongated, twisted, paddle-shaped central tail feathers. Pomarine also larger with more rounded appearance. **Status** Common arctic migrant. **Notes** Often chases terns and gulls to force them to disgorge fish they have caught.

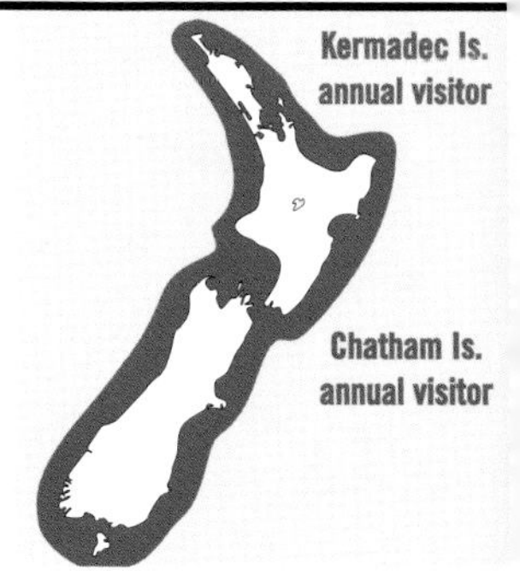

Range Breeds in higher latitudes of Northern Hemisphere. Summer visitor to New Zealand waters.
Where to see Coastal waters of New Zealand in summer, often near Red-billed Gull (p. 114) and White-fronted Tern (p. 120) colonies.

Pomarine Skua

Stercorarius pomarinus

Identification 48 cm. Medium-sized, robust, rounded skua, with similar range of plumages as Arctic Skua (above). Most common plumage phase blackish-brown above and white below with yellowish flush to nape. Immatures have distinctive barred rump and undertail coverts, and rump appears paler than rest of upperparts. Two conspicuous white patches at base of primaries, rest of underwing dark. Breeding adults have two elongated central tail feathers which are paddle-shaped and twisted. **Similar species** Arctic Skua (above), but Pomarine noticeably larger, more rounded and with two white patches at base of primaries. Adults also have paddle-shaped rather than pointed central tail feathers and immatures have more heavily barred rumps and undertail coverts. **Status** Uncommon arctic migrant. **Notes 1.** May follow vessels. **2.** Harries terns, gulls and sometimes waders. **3.** More oceanic than Arctic Skua.

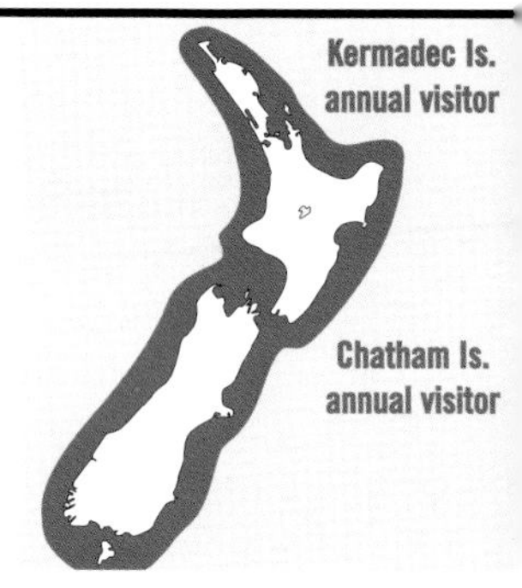

Range Breeds in Arctic. A regular summer visitor to New Zealand waters.
Where to see Coastal and oceanic waters around New Zealand.

Arctic Skua

Pomarine Skua

Black-backed Gull / Karoro

Larus dominicanus

Identification 60 cm. Large, white-bodied gull with black back and wings tipped with white. Bill yellow with red spot on lower mandible. Legs are olive-yellow. Juveniles dull mottled brown. Flight slow steady strokes interspersed with long glides. **Similar species** No similar species in New Zealand, but juveniles could be confused with the much larger, darker Brown Skua (p. 110), but Black-backed Gull lacks the white wing patches of skuas. **Status** Abundant native. **Notes 1.** An enthusiastic scavenger around rubbish dumps. **2.** Often feeds on carrion, along beaches and inland on farms. **3.** Numbers have increased as a result of human activities.

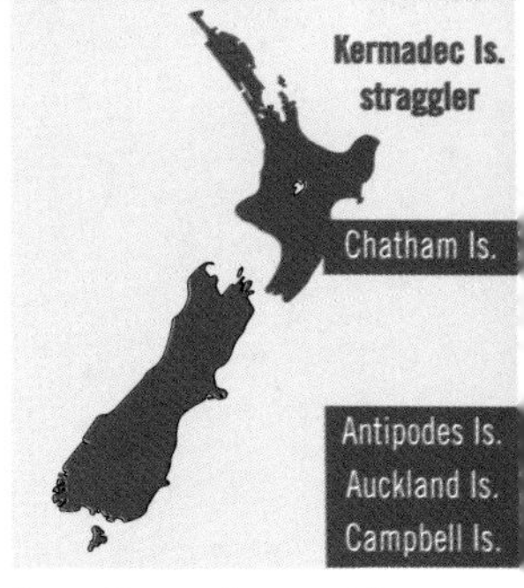

Range Circumpolar around the Antarctic. In New Zealand widespread throughout the North, South and Stewart Islands in coastal and inland waters. Also at Chatham, Auckland, Campbell and Antipodes Islands. Seldom ranges far offshore.
Where to see In most coastal and lakeside areas.

Red-billed Gull / Tarapunga

Larus novaehollandiae

Identification 37 cm. Smallish, white gull, with silver-grey back and wings. Wingtips black with small white patch. Conspicuous red bill and legs. Juvenile has some brown mottling on upperwings, pinkish bill with dark tip and dull pinkish legs. **Similar species** Adult Black-billed Gull (p. 116) could be confused with juvenile Red-billed, but Red-billed has mainly black wingtips and bill is more robust. **Status** Abundant native. **Notes 1.** New Zealand subspecies is *L. n. scopulinus*, although some authorities consider it to be a full species, *L. scopulinus*. **2.** An enthusiastic scavenger whose numbers have greatly increased with growth of the human population.

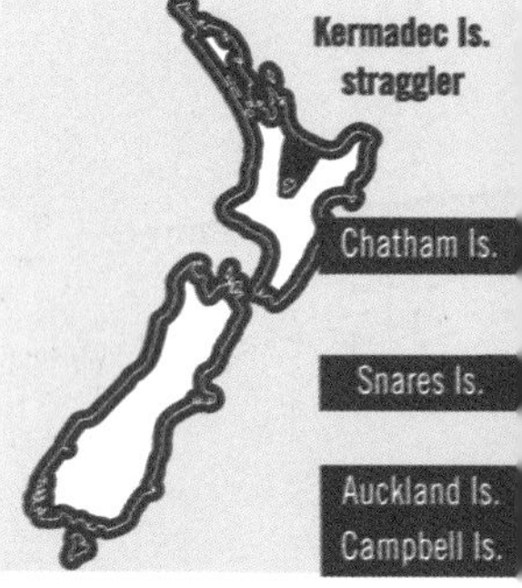

Range Occurs in subantarctic from Africa to Chatham Islands. Throughout New Zealand mainland and subantarctic islands. Also in Australia and New Caledonia.
Where to see Coastal areas throughout mainland New Zealand and at Lake Rotorua.

Black-backed Gull / Karoro adult and juvenile (inset)

Red-billed Gull / Tarapunga adult and juvenile (inset)

Black-billed Gull / Tarapunga

Larus bulleri

Identification 37 cm. Very pale gull, white with pale grey wings and white wingtips thinly edged with black. Slender, black bill and black legs. Juvenile has some mottling on upperwings, pinkish bill with dark tip and pinkish legs. **Similar species** Juvenile Red-billed Gull (p. 114), but Black-billed has less black on wingtips and more slender bill. **Status** Common endemic. **Notes 1.** Much more an inland species than Red-billed Gull. **2.** In the south, often follows the plough.

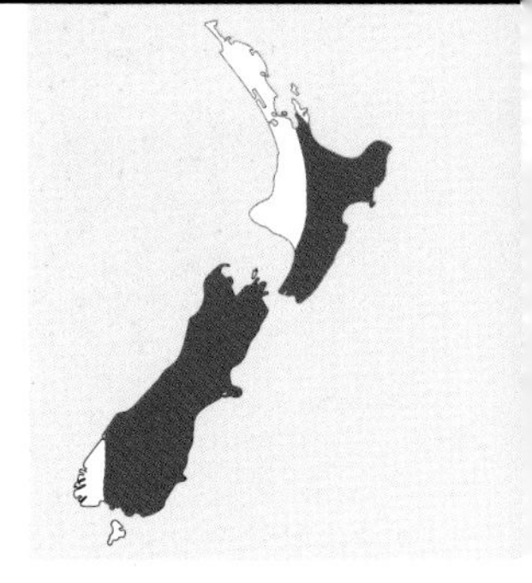

Range Mainly inland waters of the North and South Islands. Also a few coastal colonies. Most move to coasts in winter.
Where to see Around Lake Rotorua and the South Island lakes.

White-winged Black Tern

Chlidonias leucopterus

Identification 22 cm. Small tern with black body and pale grey wings. Bill and legs reddish-black. In eclipse (non-breeding) plumage, body white, upperparts pale grey with diagnostic black patch covering ears, occasionally extending to nape. Flight light and buoyant. **Similar species** Little Tern (p. 124), but White-winged Black can be distinguished by its black ear patch. **Status** Uncommon Eurasian migrant. **Notes 1.** Occasionally rests on power lines, atypical tern behaviour. **2.** Has attempted to breed in New Zealand.

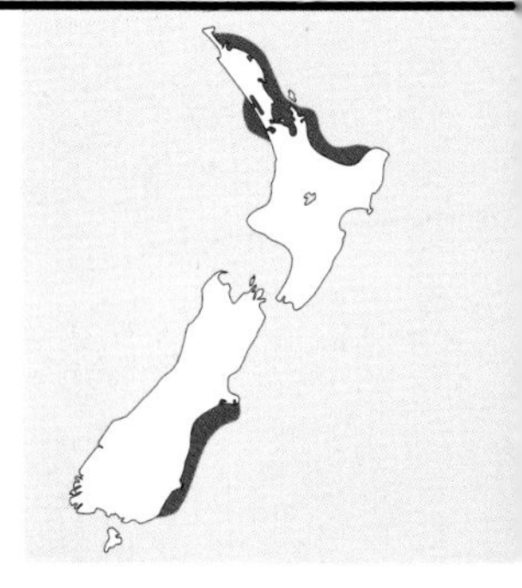

Range Uncommon migrant to sheltered estuaries and harbours from Northland to Foveaux Strait. Rarely over-winters.
Where to see Sheltered coastal waters around New Zealand. Sometimes seen over coastal lagoons.

© Brian Parkinson

Black-billed Gull / Tarapunga

© Brian Chudleigh

White-winged Black Tern (eclipse plumage)

Gull-billed Tern

Gelochelidon nilotica

Identification 40 cm. Large, stocky, pale tern with short, stout, black bill and long, black legs. Black cap, which in eclipse phase is reduced to black patch over ear coverts. **Similar species** Caspian Tern (p. 120), but Gull-billed has a black bill, not red. **Status** Rare subtropical visitor. **Notes 1.** Subspecies which visits New Zealand is probably *G. n. macrotarsa*. **2.** Sometimes seen over inland farms.

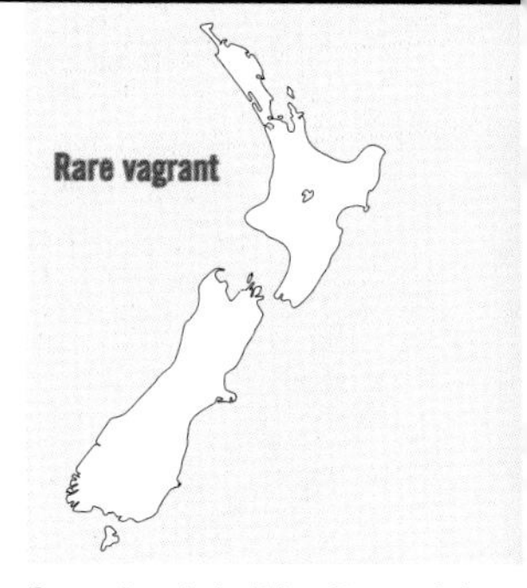

Range Breeds in Africa, Europe, Asia and the Americas. Very occasional visitor to New Zealand coastal waters. Rarely over-winters.
Where to see Sheltered coastal waters south to Cook Strait.

Black-fronted Tern / Tara-piroe

Sterna albostriata

Identification 29 cm. Small blue-grey tern with conspicuous white rump, greyish tail and black cap. Bill and legs orange. **Similar species** White-fronted Tern (p. 120), but the Black-fronted is smaller, greyer, has orange bill and lacks white forehead band. **Status** Locally common endemic. **Notes** Vulnerable to introduced predators at riverbed colonies.

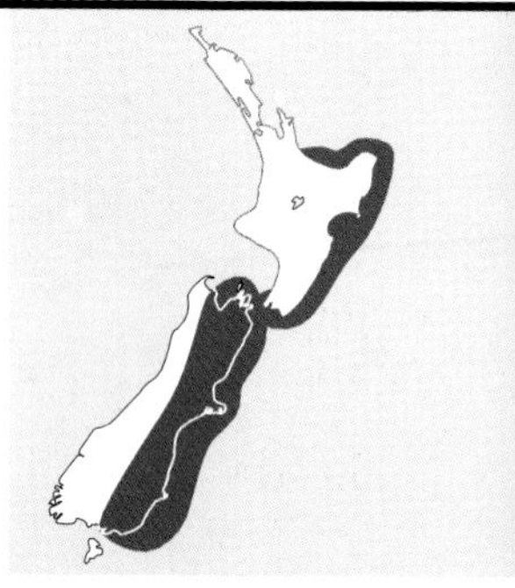

Range A riverbed breeder of the South Island, east of the Southern Alps. In winter migrates to east coast of South Island, Cook Strait and the Wellington coastline, and small flocks regularly reach the Bay of Plenty.
Where to see South Island riverbeds in summer. In winter at Farewell Spit and Waikanae Estuary, also Cook Strait and coastal waters of North and South Islands south of East Cape.

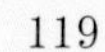

Gull-billed Tern

Black-fronted Tern / Tara-piroe

Caspian Tern / Tara-nui

Hydroprogne caspia

Identification 51 cm. Very large tern with white body and grey wings. Prominent red bill, cap black in breeding season, becoming grizzled during non-breeding periods. Juvenile has duller bill and brownish cap. **Similar species** Large size and conspicuous red bill make it difficult to confuse Caspian Tern with other terns. **Status** Reasonably common native. **Notes 1.** Nests in large colonies on sand and shell banks, roosts in small flocks. **3.** Some northward movement of more southerly breeding birds in autumn and winter.

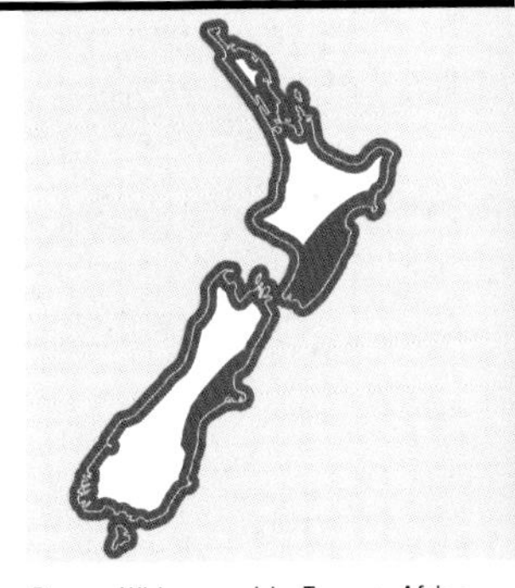

Range Widespread in Europe, Africa, Asia and North America. In New Zealand, widely distributed around the coastline.
Where to see Sheltered coastal waters, harbours and estuaries of North and South Islands. Occasionally on lakes and some riverbeds.

White-fronted Tern / Tara

Sterna striata

Identification 40 cm. Medium-sized, very white tern with prominent black cap, and pale grey wings with dark line on outer primary. Black legs and bill. Distinctive white band of feathers between cap and bill gives this species its name. Immature has brown markings on upperparts, dark mottling on front edge of upperwing and less distinctive cap. **Similar species** Black-fronted Tern (p. 118) but the Black-fronted is smaller and greyer with no white band on forehead and has an orange bill. **Status** Abundant endemic. **Notes 1.** Forms large, noisy flocks over shoaling fish such as kahawai. **2.** In summer, often harried by Arctic Skuas (p. 112).

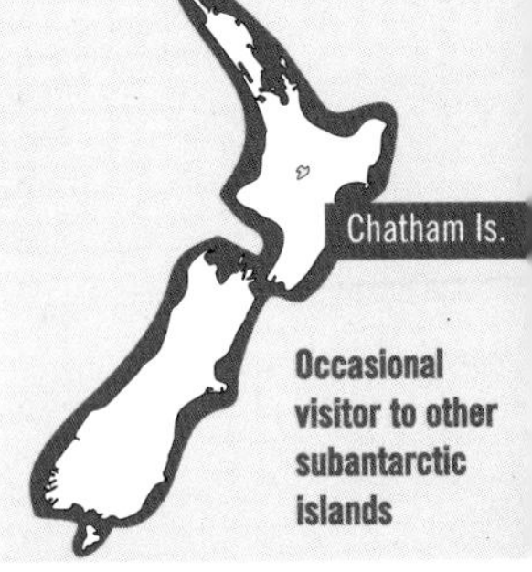

Range Widespread around mainland coastlines and on Auckland and Chatham Islands. Many birds including most juveniles winter in Australian waters.
Where to see Around New Zealand coastline, especially inshore waters.

Caspian Tern / Tara-nui

White-fronted Tern / Tara

Sooty Tern

Sterna fuscata

Identification 45 cm. Medium-sized tern with conspicuous black wings and cap which contrast sharply with white forehead and white underparts. Long, forked tail conspicuous in flight. Bill and legs black. **Similar species 1.** The black upperparts distinguish Sooty Tern from other terns in New Zealand waters. **2.** A very rare vagrant to New Zealand, the Bridled Tern is similar, however Bridled is smaller, browner and white forehead patch extends behind eye. **Status** Locally common native. **Notes** A pelagic species often encountered far from land.

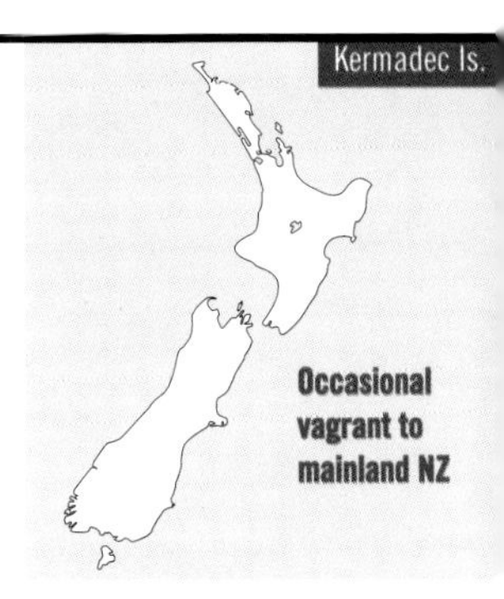

Range Widespread tropical and subtropical distribution in Atlantic, Indian and Pacific Oceans. In New Zealand region, breeds at the Kermadec Islands. A rare visitor off Northland in autumn and winter after gales.
Where to see Around Kermadec Islands and rarely in waters north-east of Northland.

Antarctic Tern / Tara

Sterna vittata

Identification 36 cm. Medium-sized tern with grey body and wings, white rump and tail, and distinctive black cap which reaches down to bill. Bill and feet red, becoming brighter in breeding season. **Similar species** Could be confused with White-fronted Tern (p. 120) around southern Stewart Island and Auckland Islands where both coexist, but Antarctic Tern is smaller and greyer, has a red bill and lacks white forehead stripe. **Status** Locally common native. **Notes 1.** The New Zealand subspecies is *S. v. bethunei*. **2.** At Stewart Island present in coastal waters all year.

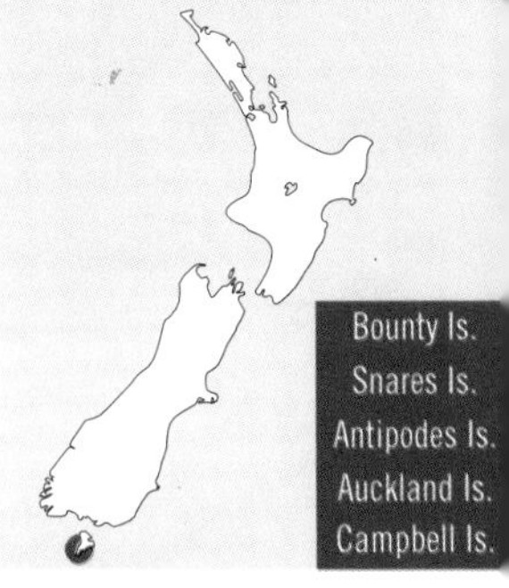

Range Circumpolar subantarctic. In New Zealand subantarctic occurs on islands around southern Stewart Island, Snares, Campbell, Auckland, Antipodes and Bounty Islands. Not known to reach New Zealand mainland.
Where to see In waters to the south of Stewart Island and around the subantarctic islands.

Sooty Tern

Antarctic Tern / Tara

Fairy Tern / Tara-iti

Sterna nereis

Identification 25 cm. Very small, white body, pale grey wings and black cap with white forehead. Orange bill and feet. In non-breeding plumage black cap smaller and duller, bill dusky orange with dark tip, legs dull orange. Immature has dark bill and feet and dark leading edge to wing. Rapid flight, often hovers. **Similar species 1.** Little Tern (below), but Fairy lacks black-tipped bill, black line from cap through eye to bill and dark primaries. **2.** Apart from Little Tern, Fairy Tern is smaller than most other terns. **Status** Rare native. **Notes 1.** Once more widespread in New Zealand, but now one of the rarest shore birds. **2.** Department of Conservation has formulated a recovery plan for this species. **3.** New Zealand subspecies is *S. n. davisae*.

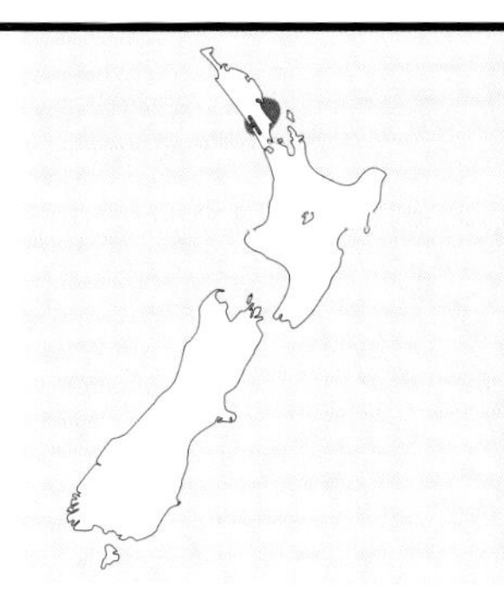

Range Central Northland beaches and sandspits at Mangawhai, Waipu and Kaipara.
Where to see Mangawhai and Waipu Estuaries on the Northland east coast, Tapora and Papakanui spit on the Kaipara Harbour.

Little Tern

Sterna albifrons

Identification 25 cm. Very small tern, grey above and white below with black cap and yellow bill tipped with black. Breeding adult has white forehead and dark line extending from cap through eye to base of bill. Outer 2–3 primaries black. Non-breeding adult (as seen in New Zealand during southern summer) less distinct cap, which does not extend to base of bill. Bill dull yellow with dark tip, legs dull yellow. Immature similar to non-breeding adult but has dark bill. Rapid flight, often hovers. **Similar species** Fairy Tern (above), but Little has yellower bill with black tip, black line from cap through eye to base of bill, darker grey upperparts and dark outer primaries. Non-breeding adult Fairy and Little Terns very similar, but look for steeper forehead, paler grey upperparts and lack of dark outer primaries in Fairy Tern. **Status** Uncommon Asian migrant. **Notes** Often very vocal when feeding.

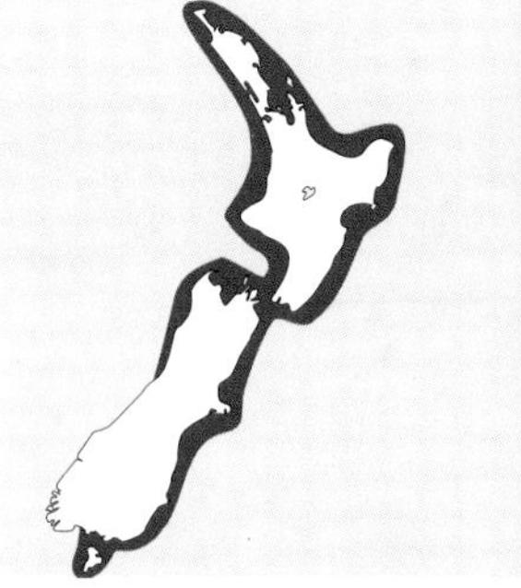

Range Migrant to New Zealand from eastern Asia where they breed. Some may come from Australia where they also breed. Not known to breed in New Zealand. Singles and small flocks visit New Zealand every summer, especially the northern harbours, e.g. Kaipara, Manukau, Firth of Thames.
Where to see Northern North Island harbours and estuaries especially Kaipara Harbour and Firth of Thames during summer.

© Don Hadden

Fairy Tern / Tara-iti

© Brian Chudleigh

Little Tern

Crested Tern

Sterna bergii

Identification 47 cm. Large tern with distinctive shaggy black cap, lemon yellow bill and black legs and feet. **Similar species** Shaggy black cap distinguishes Crested from other New Zealand terns. **Status** Rare tropical vagrant.

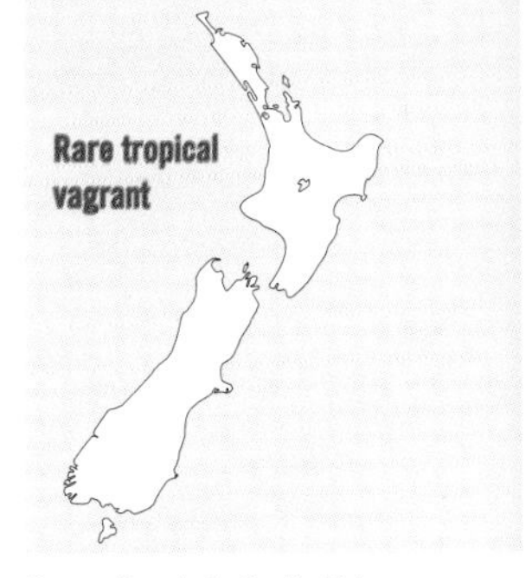

Range Breeds in South Africa, Australia and on some Indian Ocean islands. In New Zealand region, rare visitor to sheltered coastal waters. **Where to see** Rarely, in sheltered coastal waters of New Zealand.

Common Tern

Sterna hirundo

Identification 35 cm. A smallish tern, usually only encountered in New Zealand in winter when in non-breeding plumage. In non-breeding plumage, forehead white and black cap extends from in front of eye to nape. The underparts white, greyer on stomach. Upperparts grey apart from faint black shoulder bar. Bill black, feet reddish-black. In flight, tail deeply forked and outer dark grey primaries contrast with paler primaries. **Similar species** White-fronted Tern (p. 120), but the Common is smaller and greyer, bill is shorter and finer and legs are longer. It also has faster wingbeats and more buoyant flight. **Status** Rare arctic migrant. **Notes** The subspecies that visits New Zealand is *S. h. longipennis*.

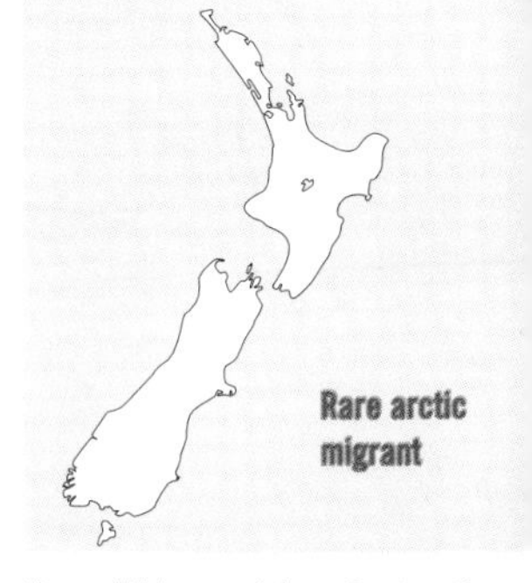

Range Widespread throughout northern Europe, north-east Asia and North America. Rare vagrant to sheltered coastal waters of the North Island. **Where to see** Sheltered coastal waters especially in Bay of Plenty.

Crested Tern

Common Tern (eclipse plumage)

Common Noddy

Anous stolidus

Identification 39 cm. Medium-sized, dark brown tern, has broad tail with shallow notch. Pale forehead and crown, centre of underwings silvery grey. Distinctive buoyant flight. **Similar species** White-capped Noddy (below), but Common Noddy larger, browner and has pale centres to underwings. Tail also has shallower notch. **Status** Rare native at Kermadec Islands. **Notes 1.** At Kermadecs nests on rocky ledges as trees lacking (usually nests in trees). **2.** Abundant in tropical Pacific where it forms large feeding flocks with White Terns (p. 130) and boobies.

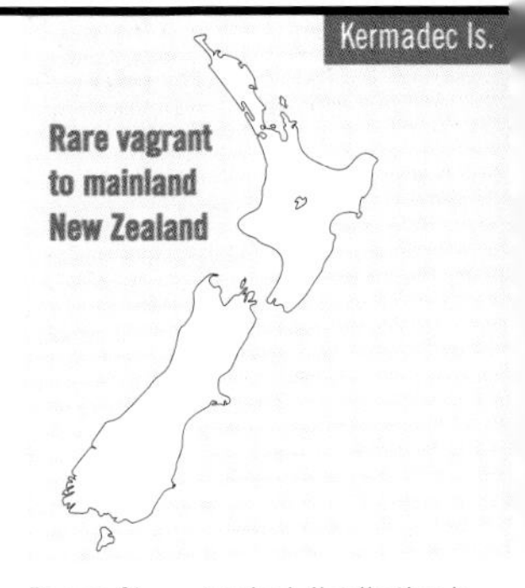

Range Circumtropical distribution in Atlantic, Indian and Pacific Oceans. In New Zealand region, breeds in small numbers at Curtis Island, Kermadec group. Very rare vagrant to mainland New Zealand waters.
Where to see Seas around southern Kermadec Islands.

White-capped Noddy

Anous tenuirostris

Identification 37 cm. Medium-sized dark tern with broad, slightly notched tail, silvery white forehead and crown, and black underwings. Distinctive buoyant flight. **Similar species** Common Noddy (above), but White-capped Noddy smaller and blacker with black underwings. **Status** Locally common native (at Kermadec Islands). **Notes** At Kermadecs usually nests in trees, but where these are lacking nests on rocky ledges.

Range Circumtropical and subtropical in Atlantic, Indian and Pacific Oceans. In New Zealand region, breeds at the Kermadec Islands. Rare vagrant to mainland New Zealand waters.
Where to see Rare vagrant to northern New Zealand. Common around Kermadec Islands.

Common Noddy

White-capped Noddy

Grey Ternlet

Procelsterna cerulea

Identification 28 cm. Small, pale blue-grey tern with conspicuous eye and pale head and neck contrasting with darker wings which have white trailing edges. Bill black. At rest, long, black legs and yellow-webbed, black feet are visible. Distinctive buoyant flight. **Similar species 1.** The absence of a black cap distinguishes this tern from most other terns in New Zealand waters. **2.** White Tern (below), but Grey Ternlet smaller and greyer. **Status** Locally common native. **Notes 1.** Subspecies in New Zealand and north-eastern Australia is *P. c. albivitta*. **2.** May feed in large flocks.

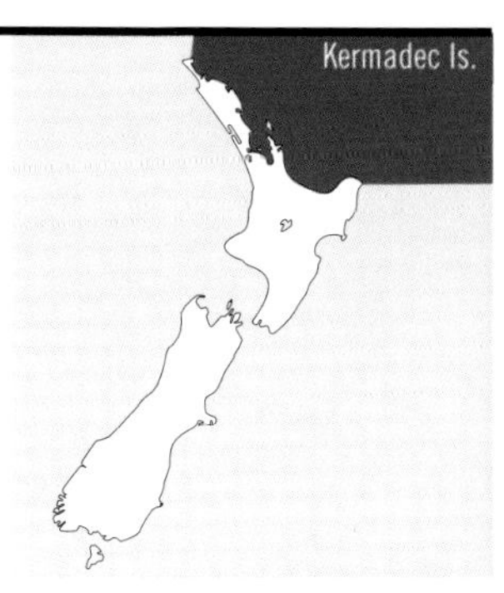

Range Widespread in tropical and subtropical Pacific. In New Zealand region, breeds in large numbers at the Kermadec Islands and apparently breeds in small numbers on rock stacks at the Three Kings and off the Northland and Bay of Plenty coasts. **Where to see** Waters north of the North Island, and outer waters of Northland, Hauraki Gulf and Bay of Plenty.

White Tern

Gygis alba

Identification 31 cm. Small, pure white tern with black ring of feathers around eye, which makes eye appear larger. Wings and short, forked tail almost translucent. Bluish-black bill, legs black, feet grey-blue with yellow webs. **Similar species** Only small, all-white tern. **Status** Rare native. **Notes 1.** South-west Pacific subspecies is *G. a. royana*. **2.** A surface feeder which does not dive. Follows shoals of small fish or crustaceans. **3.** Builds no nest; single egg laid in a depression on a bare branch.

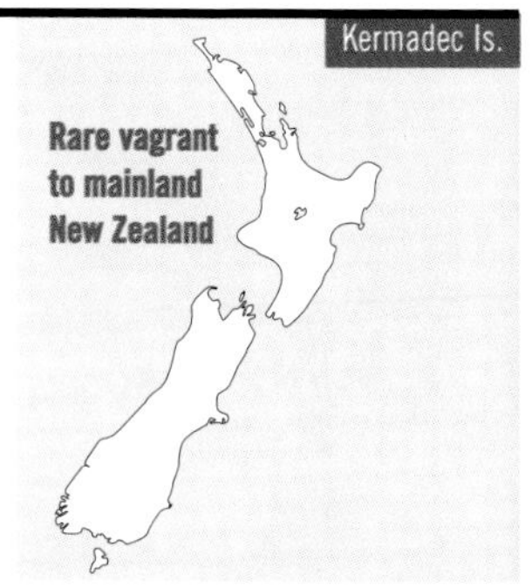

Range Circumequatorial distribution in Atlantic, Indian and Pacific Oceans. In New Zealand region, breeds on the Kermadec Islands. Also breeds on Lord Howe and Norfolk Islands. **Where to see** Around the Kermadec Islands. Rare straggler to mainland New Zealand.

Grey Ternlet

White Tern

Index

Notes